ISO 14000

Understanding the Environmental Standards

W. M. von Zharen

Government Institutes, Inc.
Rockville, Maryland

Government Institutes, Inc., 4 Research Place, Suite 200, Rockville, Maryland 20850

99 98 97 96 5 4 3 2

Library of Congress Cataloging-in-Publication Data

Von Zharen, W.M.

ISO 14000: Understanding the environmental standards / by W. M. Von Zharen.
206 p. cm.
Includes bibliographical references and index.
ISBN: 0-86587-510-3
1. ISO 14000 Series Standards I. Title
TS155.7.V66 1995
658.4 ' 08--dc20

95-46966
CIP

Printed in the United States of America

This book is dedicated to Margee, Sev, and Penny.

Summary of Contents

Table of Contents

Foreword

If you have an interest in environmental management systems and compliance, no doubt you have or will read many books on the subject. This book should be at the top of your reading list because Dr. von Zharen brings great insight and a balanced perspective to a complex subject. She provides an authoritative, accurate source of information on a highly dynamic topic, including pertinent information ranging from the development process of the environmental standards through the content and interpretation of similarities with other management system standards, both quality and environmentally related. She also addresses implementation guidance including assistance on documentation and, critically, the legal ramifications in a practical, pragmatic way with emphasis on an integrated management systems approach rather than on a duplicative approach for environmental systems separate from other existing or proposed management systems (like ISO 9000).

Finally, guidance in the determination of whether to pursue certification of your environmental management system is provided in an honest, forthright manner that does not attempt to "sell" the reader on any particular course of action. This is followed by an excellent review of the certification process and key factors to consider when selecting a partner in the process.

Altogether, this book serves as an admirable resource for everything related to ISO 14000 and environmental management systems, from creation of the standards through implementation and continuous improvement of your organization's environmental performance.

Because the only certainty is the continued changing face of standardization and the impact it will have on businesses everywhere, Dr. von Zharen provides a vital and outstanding perspective and an integrated approach that should not be overlooked by any organization considering implementation of ISO 14000.

Elizabeth A. Potts
President, ABS Quality Evaluations
An Affiliate of the American Bureau of Shipping

Preface

When teaching Business Law, I invariably include a section on the ISO 9000 and 14000 series, summing up the underlying strategy of the ISO 9000 system thus: "Say what you do; do what you say; and then prove it." Without fail, this summation triggers a few flailing hands and the question is posed to me: "So ISO 9000 has nothing to do with quality then? Only *consistency?*" I nod.

By the same token, the ISO 14000 series has nothing to do with ensuring that a product or service is the most environmentally sound one. What it does, however, is provide a systematic approach to continual improvement of an environmental management system.

The intellectual reception of the ISO 14000 series has been various: from the bane of existence for some organizations because of the perceived immediate increased costs—an expensive refuge for number-crunchers and a coercive business practice—to a living document for other organizations which addresses the concept of improved quality of the environment. No matter the moniker, it is gaining increasing attention worldwide as an ambitious attempt to create a management system in which environmental concerns permeate the business strategy.

Seminars on this topic are crowded with environmental managers and CEOs, but oftentimes these participants leave without a clear understanding of how to address the many facets of the ISO 14000 series. This book is intended to fill that void, to answer the questions often posed regarding the varied aspects of the series, and to discern just what the ISO 14000 series is, why an organization should care, how it may be implemented, and its potential pitfalls. To meet these objectives, this book begins by describing the International Standards Organization as well as its history. It also delineates the steps involved in creating an international standard. Chapter Two serves to introduce readers to the field and scope of the ISO 14000 series and highlights other environmental standard systems. Chapter Three is the first of the "practical" chapters which reviews each section of ISO 14004, explains the jargon and nomenclatures, and "walks" the reader through the guidelines. Chapter Four reviews each section of ISO 14001 and discusses the various requirements for certification or self-declaration to this specification document. Chapter Five examines each section of the auditing documents, ISO 14010, 14011, and 14012. Chapter Six provides the reader with an overview of implementing ISO 14001 and development of an environmental management system. It identifies the basis for the system, identifies organizational conundrums, and provides guidance for the process of implementation. Chapter Seven discusses documentation requirements, document control, and records. Chapter Eight explores the world of certification bodies, registrars, and the certification process. It pinpoints what to look for in choosing a registrar. Chapter Nine addresses audits and discusses why there should be an audit, the types of certification audits, as well as audit procedures. Chapter Ten highlights various legal issues involved in auditing and focuses on the protection of audits as "privileged" business information and mitigation of damages. The final Chapter includes a cautionary approach to implementing an EMS.

Author's Note

When writing this book, it was the intention of the author to provide interested parties as quickly as possible salient information on the more important aspects of those ISO 14000 series documents that have become Draft International Standards (DIS): ISO 14004, Environmental Management Systems—General Guidelines on Principles, Systems, and Supporting Techniques; ISO 14001, Environmental Management Systems—Specification with Guidance for Use; ISO 14010, Guidelines for Environmental Auditing—General Principles; ISO 14011, Guidelines for Environmental Auditing: Audit Procedures—Auditing of Environmental Management Systems; and ISO 14012, Guidelines for Environmental Auditing—Qualification Criteria for Environmental Auditors. The final documents may include minor changes of language. A second edition the ISO 14000 series will include any such changes, additional environmental standards that have been developed, and other practical information.

About the Author

Dr. von Zharen holds a dual appointment as graduate faculty for both the College of Geosciences and Maritime Studies and the College of Wildlife and Fisheries and is an associate professor of admiralty, corporate, and environmental law. She is a visiting professor of International Environmental Law in the Post-Graduate Law Program at the Universidad Autonoma de Guadalajara, Mexico, and at Memorial University in Newfoundland, Canada; and is Coordinator of Marine Policies and Management for the Texas Institute of Oceanography (TIO) and the University's International Maritime and Environmental Regulatory Affairs Option Program which includes cooperative international environmental management programs with Latin America, Asia, Europe, and Russia. She also serves as Legal Counsel for Environmental Affairs to the American Bureau of Shipping's Marine Services, Inc., and is a lecturer for the American Association of Petroleum Geologists. She developed the "Partners in Leadership for Sustainable Development" Program with Texas A&M University, Memorial University, Newfoundland, Canada, and the Universidad Autonoma de Guadalajara, Mexico, a program which centers on development of marine and coastal resources and marine transportation. She holds graduate degrees from the University of Florida, the University of Texas, and the University of South Carolina and has published numerous articles, chapters, and several books in the area of law, particularly marine resources law and environmental system management. She is active in numerous organizations including chair of the American Bar Association Section on Natural Resource, Energy, and Environmental Law's Marine Resource Committee, vice-chair of its International Environmental Law Committee, and serves as Proctor in the Maritime Law Association; she is also a member of the College of Texas Bar and is a Faculty Advisor to the University of Texas School of Law's Admiralty Law Conference. Recently, she was selected the "Center for Teaching Excellence Scholar."

Before joining Texas A&M, she was an attorney with Exxon Company, U.S.A., and Exxon Shipping Company and an adjunct professor of admiralty and environmental law at Texas Wesleyan University. Prior to that she lived in Scandinavia and Europe as a Fulbright Scholar and Researcher, studying primarily in the area of proxemic analysis, which she later developed for use in legal negotiation, conflict resolution, and business management strategy. She completed research in various African countries, the South Pacific, and Asia including work in Ponape with the Pacific Studies Institute.

Acknowledgments

I am grateful to the many people who assisted me with this project and would particularly like to thank my research assistant, Ms. Patricia Mentzer, who ably dedicated herself to the book and performed prodigious feats of research, editing, and proofing. Special appreciation is due to Mark A. Klugiewicz of ABS Quality Evaluations, Inc., who contributed research material for this endeavor. I also value the enthusiastic support provided to the project by the Texas A & M University's administration including my Dean, Dr. David Schmidly, and my Department Head, Dr. James McCloy. Recognition must also be given to Dr. G.K. Jacobs.

1
Development of International Standards

Too much consistency is as bad for the mind as it is for the body. Consistency is contrary to nature, contrary to life. The only completely consistent people are the dead.

Aldous Huxley, Do What You Will (1929) "Wordsworth in the Tropics"

The Quest for Universal Standards

Many businesses by their very nature must position themselves as global in scope. Concomitant with this comes the increased demand for assurances of consistency and quality to reduce risks. As a result, acceptance of international standards is increasing; the ripple effect ensures that as companies are ISO 14001 certified, their suppliers, in turn, must be certified resulting in a sound, consistent environmental management policy ultimately effectuating management of environmental risks and protection of the environment. In a global market, environmental initiatives such as the ISO 14000 series may change the way companies do business.

Dynamics of the International Standards Organization (ISO)

The International Standards Organization (ISO), based in Geneva, is an international federation of "standards bodies" from one-hundred-and-eleven nations.[1] It was founded in 1946 with the aim of promoting standardization and

---

1 Working within the same genre is the Comite Europeen de Normalization (CEN), a committee which drafts standards on behalf of the fifteen European Union (EU) countries as well as six nations belonging to the European Free Trade Association (EFTA). The CEN's procedures, for the most part, mirror those of the ISO with standards adopted on the basis of

related activities in order to facilitate international exchange of goods and services. Historically, it addressed primarily technical product or manufacturing standards; it then evolved into quality control system standards; currently, it is addressing environmental management systems.[2] The ISO develops standards in all industries except those related to electrical and electrical engineering.[3]

The ISO may bring together such disparate interests as those of manufacturers, product users, governments, the scientific community, and environmental groups. The ISO typically produces two types of standards: guidelines and specification documents. The guidelines offer direction as to what *should* be done while the specification documents detail what *shall* or must be done. From each member country, only one national body representing standardization for the nation may belong to the ISO.

The path of development for a standard is complex primarily because of the large number of entities who have varying input on its progress. Member nations are divided according to their degree of participation. Three "participation based" categories are allowed: full, correspondent, and subscriber. A *full* member is the national body "most representative of standardization in its country." Typically, this is the national standards body, a government institute, or government-sponsored organization. Currently, there are eighty-two full members.[4]A *correspondent* member may not participate actively in the technical work but is kept informed of the work of interest to it. A typical member is an organization such as a government institute in a lesser-developed country that does not have its own national standards body. Currently, there are twenty-two correspondent members.[5] A *subscriber* membership is designed to accommodate countries with relatively small economies. These members pay a reduced fee but are not allowed to attend

majority vote. Once it is decided that a standard is needed, the choice is made whether to develop it via the CEN or through the ISO. An agreement between the CEN and the ISO was signed in 1991, the so-called Vienna Agreement, in which an attempt was made to route standards development through ISO. However, the choice is still made on a case-by-case basis depending on the standard.

[2] More than 350 international standards with environmental relevance are on ISO's books.

[3] Those standards are developed by the International Electrotechnical Commission (IEC) based in Geneva.

[4] As of March, 1995.

[5] As of March, 1995.

technical meetings; they can attend other meetings only as observers. They do not have access to documents.

There are two categories of membership on technical committees (TC), each with a different degree of participation. "Participating members" ("P") have the right to vote, attend all TC meetings, and collect documents; participation of "observer members" ("O") depends on their ISO status (full, correspondent, or subscriber). Only *full* members of ISO may choose TC status—either participating status or observer status. *Correspondent* members may only have "observer membership" status which allows attendance at TC meetings as observers but does not allow voting. S*ubscriber* members may assume "observer membership" status but are not allowed to attend technical meetings; they can attend other meetings only as observers.

The process of creating a final international standard involves several steps. A brief review of these is offered:

Step One. The Proposal

When the need for an international standard has been established (*i.e.*, one of the outcomes of the Rio Conference[6] in 1992 was the realization that international environmental standards should be developed), then this new work item is submitted for vote by the members of the relevant technical committee/subcommittee (TC/SC) to determine whether to include the work item in the program. Any member country may propose a technical committee; all member countries then review this proposal. When a technical committee has been formed, its members dictate its structure. The number of a technical committee, *i.e.*, TC 207, is determined by chronology. The same is true for the numbering of documents. The ISO 14000 series was the next series to be allocated after 13000 with, in general, each set of documents in a series being numbered in increments of ten.[7] Some technical committees are comprised of subcommittees (SC); others also include working groups (WGs).

---

6 This conference produced The Rio Declaration on Environment and Development.

7 For example, in the ISO 14000 series, the auditing documents begin with 14010 and include 14011 and 14012.

Step Two. The First Draft

Once the technical committee decides to work on a standard, a working group convener is appointed. Depending on how many documents are to be produced, the TC/SC designates experts to work on drafts with each working group (WG) dedicated to a single document. Once a draft has been written, it is presented to the subcommittee for review and consensus-building. When this review phrase is completed, it is ready for submission; the working group convener presents this resolution to the technical committee or subcommittee to elevate the document to a "committee draft" (CD).

Step Three. Draft International Standard (DIS)

The first committee draft is registered perfunctorily by the Central Secretariat and the document is circulated for comments, ballot , or both, by the members of the TC/SC. There may be extensive re-drafting until a consensus is reached. The final draft is submitted to the ISO Technical Committee, such as ISO TC 207, as a draft international standard (DIS). The DIS then goes through translation into French.

Step Four. Approval

The DIS is circulated to all member bodies for voting and comment. Such members have six months in which to submit their decision. The DIS becomes an international standard only if a two-thirds majority of the participating ("P") members of the TC/SC vote for approval and not more than one-quarter of ISO full members vote negatively; this step includes comment resolution. A DIS which fails to achieve the requisite vote is returned to the originating TC/SC for further study. A revised document will then be circulated for voting and comment. Once the six-month period of ISO member body voting is complete, there is a further two-month ISO member body voting period on the "final" version.

Step Five. Publication

Once the DIS is approved, a final text is submitted to the Central Secretariat who publishes the standard. However, that is not the end of the cycle: every five years the TC/SC responsible for the standards reviews it and a majority of the participating members decide on possible reconfirmation, revisions, or withdrawal of the standard.

International Standards Organization's ISO 9000 Series

Introduction

The harmonization of standards which creates a common denominator for businesses is a far-reaching goal. One of the latest and most successful attempts is the International Standards Organization's 9000-9004 series of documents and standards on quality assurance. The term "quality," however, is not based on the traditional sense that a product or service is of higher or better quality. In fact, the ISO 9000 series, simply stated, maintains that a company "says what it does; does what it says; and proves it." Thus, the ISO 9000 series may be more appropriately termed a "consistency system" rather than a quality system.

Background

The ISO 9000 series is an international set of standards written by the ISO/Technical Committee 176 (ISO/TC 176), a world-wide delegation consisting of standards committees. National associations participating in the ISO/TC 176 include: the American National Standards Institute (ANSI); the British Standards Institute (BSI); the Association Francaise de Normalization (AFNOR); the Nederlands Normalisatie Institüt (NNI); and the Standards Council of Canada (SCC). Individual member countries may be represented by their national standards association. In the United States, the series is known as the ANSI/ASQC (American National Standards Institute/American Society of Quality Control) Q90-Q94 Series.

The core standards are not industry specific but rather provide a general framework to which any industry may conform. The ISO 9000 series includes five documents. The title of each is descriptive:

- *ISO 9000: Quality Management and Quality Assurance Standards: Guidelines for Selection and Use;*

- *ISO 9001: Quality Systems: Model for Quality Assurance in Design/Development, Production, Installation, and Servicing;*
- *ISO 9002: Quality Systems: Model for Quality Assurance in Production and Installation;*
- *ISO 9003: Quality Systems: Model for Quality Assurance in Final Inspection and Test;* and
- *ISO 9004: Quality Management and Quality System Elements Guidelines.*

Three documents represent distinct functional capabilities while two are meant only as guidelines:

- Three quality assurance, function-specific models -
  - (a) 9001/Q91;
  - (b) 9002/Q92; and
  - (c) 9003/Q93.
- Two declaratory/guideline documents -
  - (a) 9000/Q90; and
  - (b) 9004/Q94.

It is not possible to obtain registration to ISO 9000. That document provides guidelines exclusively. What is possible, however, is to obtain registration to one of the three core, function-specific documents: 9001, 9002, 9003.

Application

ISO 9001 is applicable in contractual settings when conformance to specified requirements is to be assured during several stages which may include design/development, production, installation, and services. ISO 9002 is applicable in contractual situations when:

- The specific requirements for the product are stated in terms of an established design or specification; and
- Confidence in product conformance can be attained by adequate demonstration of a certain supplier's capabilities in production and installation.

Under the ISO 9000 series models—9001, 9002, and 9003—the obligations of the supplier of services decreases. The ISO 9003 supplier must demonstrate a capability to inspect and test a product while ISO 9002 suppliers must also prove that the relevant manufacturing processes are capable of maintaining requirements as per design specifications. The ISO 9001 adds to this litany that the supplier must demonstrate mastery in all phases of design, development and servicing.

The ISO 9002 remains the standard most suited to the majority of companies delivering management services although 9001 and 9002 should be viewed as complementary. Additionally, because service operations may be subject to a full panoply of conventions, statutes, and codes of practice, these requirements may be incorporated into the ISO system depending on the particular adaptation to the service industry.

The essence of the ISO 9000 series rests with the commitment to this quality system beginning at the highest level of management. It is top management's responsibility to prepare a quality policy which will define:

- Corporate objectives with regard to service quality;
- The standards to which the service will conform;
- The method by which quality objectives will be achieved; and
- The identity and scope of those persons responsible for developing, implementing, and maintaining the quality policy.

The management must be committed to quality and consistency and must ensure that the policy is implemented, understood, and maintained at all levels in the organization.

Within the system, rigorous preparation and maintenance of documented quality system procedures, instructions, and means for effective implementation of the system must be evidenced. Other areas of attention include:

- Contract review;
- Design control;
- Document control;
- Purchasing including assessment of subcontractors and suppliers;
- Product identification and traceability;
- Process control;

- Inspection and testing;
- Inspection, measuring and test equipment;
- Control of non-conforming product;
- Corrective action;
- Handling, storage, packaging and delivery;
- Quality records;
- Internal quality audits;
- Training;
- Services; and
- Statistical techniques.

In addition, all Policy Statements are reviewed and amended as necessary so that changes in relevant legislation and other developments can be incorporated. Copies of the Quality Policy should be available to all personnel within the company and to all sub-contractors of the company. Adopting an ISO 9000 standard should result in reduced audits and better defined processes.

Development of the ISO 14000 Series

The impetus for the ISO 14000 series can be traced to the Global Environmental Initiative in Rio de Janeiro in 1992.[8] There, the United Nations Conference on the Environment and Development (UNCED), attended by over one-hundred countries, agreed on the need for further development of international environmental management programs. The Conference placed a worldwide focus on corporate environmental management.

Equally important was identifying the number and extent of national and regional environmental management schemes including labeling and auditing. The acceptance of ISO 9000 worldwide formed a natural paradigm from which to develop international environmental management standards. The UNCED petitioned the ISO to consider developing the standards. The ISO, in turn, determined that there was, indeed, a need and responded by establishing the Strategic Advisory Group on the Environment (SAGE) in 1991 to examine

[8] The appendix in the ISO 14004 standard—General Guidelines on Principles and Supporting Techniques—is entitled "Examples of International Environmental Guiding Principles." The two examples are the "Rio Declaration on Environment and Development" and the International Chamber of Commerce's (ICC) "Business Charter for Sustainable Development."

standardization of environmental management practices. Participants included representatives from business, governments, and national standards-setting organizations as well as environmental professionals.

The SAGE considered whether such standards would:

- Promote a common approach to environmental management similar to quality management;
- Enhance an organization's ability to attain and measure environmental performance; and
- Facilitate lower trade barriers.

Upon conclusion of their study, the SAGE recommended that an ISO technical committee formally consider and produce final "consensus" standards. Thus, in January, 1993, the Technical Committee 207 (TC 207) was born. The number, 207, was chosen simply because it fell within the sequence of numbers for technical committees.[9] Canada was awarded the secretariat for TC 207 and the inaugural plenary session was held in Toronto, Canada, on 2-3 June, 1993; over two-hundred delegates from over thirty countries and organizations were in attendance. The largest block of countries with voting rights is from Europe. The TC 207 framework also includes coordination and cooperation with international and regional governmental bodies including the European Union (EU), the General Agreement on Tariffs and Trade (GATT), and the Organization for Economic Cooperation and Development (OECD). TC 207 meets annually to review the progress of its subcommittees.

The TC 207 consists of six subcommittees, one for each focus area, and a working group:

- SC1: Environmental Management Systems (Secretariat: U.K.; administered by the British Standards Organization)

---

[9] Numbering of documents is based on ISO 9000 documentation procedures. Each document in a subcommittee or working group is assigned a chronological number regardless of its content with every level having its own chronology number. Thus, a new draft of an old document has a notation that supersedes a previous chronology number. A number such as ISO/TC 207/SC2/N50 would indicate: ISO - a standard; 207 - draft being developed by technical committee 207 which is coming out of; SC2 - subcommittee 2; and N50 - it is the 50th document the subcommittee has considered.

- SC2: Environmental Auditing and Related Environmental Investigations (Secretariat: Netherlands; administered by the Nederlands Normalisatie-Institüt)
- SC3: Environmental Labeling (Secretariat: Australia; administered by Standards Australia)
- SC4: Environmental Performance Evaluation (Secretariat: US; administered by the American National Standards Institute)
- SC5: Life-Cycle Assessment (Secretariat: France; administered by Association Francaise de Normalisation)
- SC6: Terms and Definitions (Secretariat: Norway; administered by Norges Standardiseringsforbund)
- WG1: Environmental Aspects in Product Standards (Secretariat: Germany; administered by the Deutsches Institut für Normünge. V.)

Using the ISO 9000 series as a guide, the committees developed management standards that would allow an organization to check whether it has systems in place to prevent or detect environmental violations. The areas include:

- Environmental management systems;
- Environmental auditing;
- Environmental labeling;
- Environmental performance evaluations;
- Life-cycle assessment; and
- Product standards.

The TC 207 was also given the mandate to develop standards with these features:

- Guidelines for performance indicators;
- An appraisal of relative environmental performance;
- Performance orientation; and
- Objectivity, measurability, repeatability, and user-friendliness.

The TC 207 was specifically instructed to exclude the following work items because they were either inappropriate or being dealt with by other technical committees:

- Setting environmental performance levels;
- Test methods for pollutants; and
- Standardization of products.

The development of such standards certainly is global: Representatives from fifty-seven countries,[10] including the United Kingdom, the Netherlands, Australia, France, Norway, and the United States, assisted with the draft standards[11] and fifteen more had observer status.[12] Japan and South Korea rivaled the United States for the largest delegations. Representatives from industry, standards organizations, trade associations, governments, and environmental organizations were included. The American National Standards Institute (ANSI), a private sector-sponsored clearing-house for voluntary standards, is the U.S. representative to the ISO. "Borderless bodies" such as the United Nations, environmental groups, and other ISO TCs have representatives.[13]

ANSI developed a parallel body to TC 207, the United States Technical Advisory Group (US TAG) to "have the opportunity to review, comment, and recommend how the U.S. will vote on all new work item proposals."[14]Sub-technical advisory groups (SubTAGs or STs) were formed for each of the international subcommittees (SCs) who, in turn, developed parallel working groups (WGs). Each has interest areas corresponding to TC 207 subcommittees

10 As of March, 1995.

11 As of March, 1995, participating members included: Argentina, Australia, Belgium, Brazil, Canada, Chile, China, Colombia, Cuba, Czechoslovakia, Denmark, Finland, France, Germany, India, Indonesia, Ireland, Israel, Italy, Jamaica, Japan, North Korea, South Korea, Malaysia, the Netherlands, New Zealand, Norway, Philippines, the Russian Federation, Singapore, South Africa, Spain, Sweden, Switzerland, Tanzania, Thailand, Turkey, Trinidad, United Kingdom, Uruguay, the United States, and Venezuela. (Forty-two)

12 As of March, 1995, observing members included: Algeria, Egypt, Hong Kong, Iceland, Lithuania, Libyan A'ab, Poland, Portugal, Slovakia, Sri Lanka, Ukraine, Vietnam, Yugoslavia (former), and Zimbabwe. Barbados has a representative as a "corespondent member." (Fifteen)

13 As of March, 1995, the number of such organizations stood at twenty-two.

14 U.S. TAG Operating Procedures.

and their working groups. Two additional groups were appointed to administer the SubTAGs: The American Society for Quality Control (ASQC)[15] and the American Society for Testing and Materials (ASTM).[16] There are more over three-hundred members in the US TAG including representatives from government organizations (i.e., Department of Defense, the Environmental Protection Agency, and the Department of Energy); industry (primarily, Fortune 500 companies); consultants; registrars/auditors/standards bodies; and public interest groups.

The U.S. experts for membership in the international working groups under TC 207 must be voted upon by the full U.S. TAG voting membership. ANSI's criteria for selection includes, among others, company or sponsor financial support, frequency of attendance at TAG meetings, participation in task groups, technical credentials, and personal attributes.

In order to finalizing current DIS status of certain ISO 14000 series documents, delegates sorted through comments on ISO 14004, "Environmental Management General Guidelines on Principles, Systems, and Supporting Techniques," and 14001, "Environmental Management System Specification," from U.S. industry, government, and environmental groups and responded to objections. The U.S. then approved the elevation of ISO 14004 and ISO 14001 to Draft International Standard status.

Commonalties/Differences between ISO 9000 and ISO 14000

Both the ISO 9000 and the ISO 14000 series focus on general management issues with an emphasis on systems. The concepts of ISO 14001 mirror those of ISO 9001: a requirement for a policy statement; top level management commitment; document control; employee training; corrective action; management review; and continual improvement. The ISO 9000 series requires an organization to incorporate all the relative duties including health and safety and the environment in the description of the quality assurance management system. In the traditional sense, however, these areas are not relative to "quality" and, as such, are sometimes overlooked and remain undocumented. The ISO 14000 series fills this gap and ensures environmental aspects of a worker's job are specified.

---

15 They work with SubTAGs 1 and 2.

16 They work with all other SubTAGs.

Significant differences are found. ISO 14001 includes sections on specific policy requirements; environmental aspect identification; objectives and targets setting; and mandatory compliance with applicable laws. The philosophy of the ISO 14000 series proposes a different challenge than ISO 9000. The ISO 14000 series continually emphasizes the need to recognize the importance of environmental protection and a commitment to such protection. Phrases in a policy manual such as "our goal is to embody life-cycle thinking" may necessitate re-evaluation of the traditional business strategy. The series' key principles require recognition that environmental management is among the highest corporate priorities; that a dialogue with internal and external interested parties be established; that management and employee commitment to the protection of the environment be developed with clear assignment of accountability and responsibility; that environmental strategic planning throughout the product or process life-cycle be encouraged; that appropriate and sufficient resources including employee training be provided to achieve targeted performance levels on an on-going basis; and that environmental performance be assessed against appropriate policies, objectives, and targets. The ISO 14001 standard describes the relationship between the ISO 9000 series and ISO 14001:

> *This standard shares common management system principles with the ISO 9000 series of quality system standards. Organizations may elect to use an existing management system consistent with the ISO 9000 series as a basis for its environmental management system. It should be understood, however, that the application of various elements of the management system may differ due to different purposes and different interested parties. While quality management systems deal with customer needs, environmental management systems address the needs of a broad range of interested parties and the evolving needs of society for environmental protection.*

The rationale for the new focus is sound: Unlike ISO 9000 and management strategy in general, the environmental arena is heavily regulated with tens of thousands of pages of law, some or many of which may be applicable to an organization. Non-compliance with environmental law may result in significant fines and enforcement consequences. Most importantly, however, the focus is a response to growing interest in protecting the environment through increased pressures on organizations to perform in an environmentally sound manner.

The emphasis in both management system series is on prevention rather than corrective action; a pro-active stance rather than "fire-fighting." The ISO 14000 series, however, is holistic, focusing on all media—land, water, and air—and how these are effected by an organization's direct and indirect activities; it is an enabling tool allowing organizations to incorporate environmental elements into their management system.

2
Field and Scope of Application

The best is the cheapest.
Benjamin Franklin.

Introduction

In comparatively few pages, the ISO 14000 series provides specific requirements and principles for environmental management. The standards are "systems" standards, not legal standards and are intended to harmonize, not invent standards. They are designed to help organizations manage their environmental obligations such as compliance with legal requirements; the standards do not supplant nor do they add to existing legal requirements. They do not tell organizations how to run their businesses but rather define management processes to be followed to control the impact an organization will have on the environment. The organization identifies what environmental impacts are acceptable within the legal framework.

The ISO 14000 series may apply to all types and sizes of organizations and are structured to accommodate diverse geographical, cultural, and social conditions. The standards apply to both service and manufacturing sectors. They are voluntary.

Unlike traditional piece-meal regulatory approaches to environmental protection, the ISO 14000 series simultaneously addresses all affected media. Commitment to continual improvement is required as is a commitment to prevent pollution. The views of "interested parties" are encouraged: for example, in ISO 14001, one of the key principles include establishing objectives and targets by considering, among other criteria, the views of interested parties.[17]

---

[17] ISO 14001, 4.2.3.

The ISO 14000 series standards are of two types: specification and guidance. Only ISO 14001 is a specification standard and thus, prescriptive. Further, the standards may be grouped as focusing on management systems and evaluation of an organization; or focusing on assessment of its products.

Management Systems and Organization Evaluation Standards:

- Environmental Management Systems (EMS)
  ISO 14001—Environmental Management System Specification

- Environmental Auditing (with three auditing standards)
  ISO 14010: Guidelines for Environmental Auditing—General Principles
  ISO 14011: Guidelines for Environmental Auditing: Audit Procedures—Auditing of Environmental Management Systems
  ISO 14012: Guidelines for Environmental Auditing—Qualification Criteria for Environmental Auditors

- Environmental Site Assessments
  ISO 1401X: Environmental Site Assessments

- Environmental Performance Evaluation (EPE)
  ISO 14031: Evaluation of the Environmental Performance of the Operational System and Its Relationship to the Environment

Product Evaluation Standards:

- Terms and Definitions for Self-Declaration Environmental Claims
  ISO 14021:

- Environmental Labeling
  ISO 14020: Goal and Principles of All Environmental Labeling
  ISO 14024: Environmental Labeling—Guiding Principles, Practices, and Criteria for Multiple Criteria-Based Practitioner Programs (Type 1)-- Guide for Certification Procedures
  ISO 10425: Principles of All Environmental Labeling

- Life-cycle assessment[18]
  ISO 14040: Life Cycle Assessment—Principles and Guidelines
  ISO 14041: Life Cycle Assessment—Goal and Definition/ Scope and Inventory Analysis
  ISO 14042: Life Cycle Assessment—Impact Assessment
  ISO 14043: Life Cycle Assessment—Improvement Assessment

- Product Standards (EAPS)
  ISO 14060: Guide for the Inclusion of Environmental Aspects in Product Standards

The series guidance document and definition documents include:

- Guidance: *ISO 14004—Environmental Management General Guides on Principles, Systems, and Supporting Techniques*

- Definition: *ISO 14021: Terms and Definitions for Self-Declaration Environmental Claims*

- Definition: *ISO 140XX: Terms and Definitions—Guide on the Principles for ISO/TC 207/SC6 Terminology Work*

The environmental management system standard, ISO 14001, is the only one to which organizations can be certified. The Guidance document *(ISO 14004)*, the Environmental Management Systems standard *(ISO 14001)*, and the auditing standards (*ISO 14010, ISO 14011* and *ISO 14012)* are scheduled to be operational in 1996. Although the focus of this book is on these five documents, a brief description of the others should be made, each of which is in various stages of development:[19]

---

[18] The ISO 14000 series defines life-cycle as the "cradle-to-grave" cycle of a product's life starting with the extraction of raw materials from the ground and continuing through the processing, manufacturing, and transportation steps, and ending with consumption, disposal, or recovery. The general categories of environmental impacts to consider include resource depletion, human health, and ecological consequences. The total effect of the product on the environment, direct and indirect, is emphasized.

[19] As of December, 1995, none of these had reached the Draft International Status.

Labeling: ISO 14024.[20] The standard addresses environmental claims of any sort, not simply labeling. This could include an organization's television advertisements, for example. These standards center on three types of environmental labeling:

Type I: Third-party certified environmental labeling;
Type II: Informative self-declaration claims; and
Type III: Quantified product information based on independent verification.

The standard is voluntary. A body develops the program, administers it, and awards a label. The draft standard's contents include:

Introduction (0.)
Scope (1.)
Normative References (2.)
Definitions (3.)
Guiding Principles and Practice (4.) Field of application, voluntary nature, credibility, relationship with regulations, life-cycle approach, selectivity of awardees, review of criteria, consultation, transparency, funding, international trade, conformity assessment, accessibility, supporting studies, costs and fees, confidentiality, and product function. (4.1 - 4.17)
Procedures for Setting Criteria (5.) Development of criteria, selection of product categories, market and product research, elements of the criteria, ecological criteria, product function criteria, development of criteria, final consultation, adoption and publication. (5.1 - 5.7)
Guide for Certification Procedures (6.)

One of the most difficult areas is assessing environmental "life cycles" of a product: How are these ascertained? How are they communicated to the

[20] In a similar vein, the International Chamber of Commerce has issued ten rules called the "Code on Environmental Advertising" which stresses that environmental advertising must be consistent with environmental regulations and make no claims that "impair public confidence in the efforts made by the business community to improve its ecological performance." It appends the ICC's "Code of Advertising Practice."

public? The standard must be based on scientific information that is repeatable and reproducible.

"Type 1" looks at a product's environmental attributes from which a judgment is made identifying by a single label that a product is "environmentally preferable." Typically, it addresses the national eco-labeling programs. It is somewhat similar to the labeling of Blue Angel, Nordic Swan, and Green Seal. As well, some private programs could qualify as a Type 1 program. "Type II" centers of the assertions made by manufacturers, *i.e.*, *"Product X uses recycled paper,"* for both labels and claims. "Type III" is analogous to nutritional labels: A product would list the environmental effects associated with the product. The benefit of this standard is the effective harmonization of eco-labels. The standard sets a framework against which other programs can be measured and through which conflicts can be resolved.

Environmental Performance Evaluation (EPE): ISO 14031. This standard defines the environmental performance evaluation of a management system and provides guidance for adopting the process. The proposed draft standards contain nine elements:

- Introduction;
- Scope;
- References;
- Definitions;
- General overview of EPE;
- Design and implementation guidelines;
- Communicating EPE information;
- Continual improvement of the EPE process; and
- Annexes of illustrative examples.

The standard provides guidance on regularly evaluating environmental performance against objectives and targets established by an EMS. Environmental performance is an organization's achievement in management, the relationship among the full range of its activities, and their significant environmental risks and effects. General criteria rather than specific legal limitation are the focus. The standard provides a "menu" of performance indicators from which management chooses those most appropriate to its organization's activities and policies. These indicators give management an additional performance measurement tool other than compliance and includes those which are:

- Relevant and useful;
- Practical and cost effective;
- Technically feasible;
- Objective and value neutral;
- Verifiable and reproducible; and
- Comparable overtime.

An EPE assesses performance; and the environmental audit verifies an EMS program. Unlike an environmental audit wherein an evaluation is scheduled and a sampling of data is analyzed at infrequent intervals by people independent of the audited function, an EPE is continuous with ongoing data collection and review process conducted by employees responsible for the organization's environmental performance.

Life-cycle Assessment (LCA): ISO 14040, 14041, 14042, 14043. These standards provide principles and procedures to take into account when evaluating the environmental attributes of the entire life cycle of a product. In conducting an analysis, an organization looks at impacts of its products or services at all stages, from raw material, use and disposal, as well as design through manufacture. These standards also provide for life cycle inventories. An LCA is a quantitative assessment tool; it harmonizes environmental language; and it creates common environmental understanding. Ultimately, LCAs are concerned with stewarding products. ISO 14041 provides those directly involved in LCAs with specific guidelines and requirements for formulating the goal and scope of a life-cycle assessment and inventory analysis. ISO 14042 gives three major categories to be considered in an impact assessment of an LCA: inventory classification, characterization, and valuation. ISO 14043 reviews case studies.

Product Standards: ISO 14060. The effect of a product[21] on the environment ranges from slight to significant; provisions in product standards may influence significantly the extent of these environmental effects. This standard seeks to

[21] The term "product" includes processes, services, and combinations thereof.

provide guidance on developing environmental criteria that other professional standards writers should consider when developing product standards in order to reduce environmental effects.

Definitions: ISO 14021. This standard defines key words. It is intended to create a comprehensive set of terms and definitions for the ISO 14000 series; develop a guidance document on terms and definitions to ensure consistency among other subcommittees; ensure that definitions developed by other subcommittees meet the technical ISO requirement; and develop additional definitions where appropriate.

Environmental Site Assessments: ISO 1401X. This standard is in a "working draft in idea" stage. It has not been assigned a number.

Why an Organization May Choose to Implement the ISO 14000 Series

The ISO 14000 series was developed to accommodate a wide-range of business sizes and types and, as such, varying uses of the ISO 14000 series are possible. The application of the ISO 14001, Environmental Management System Specification, may be used to achieve third-party certification, self-declaration, or second-party recognition between contracting parties and may be suitable for certain business relations. If an organization's goal is not certification, it may nonetheless use ISO 14004, the Introduction to the Guide to Environmental Management Principles, Systems, and Supporting Techniques (referred to as Guidelines), to initiate or improve its environmental management system. ISO 14011, Auditing of Environmental Management Systems, may be integrated within the overall formulation of the Environmental Management System. The type of use of the series depends on factors such as an organization's policy; the size of the organization; the degree of systematic management already in place; and an overall assessment of the advantages and disadvantages of implementation.

The series has been labeled a "green passport" leading to a fast "trust track" between an organization and its regulatory body. An EMS system effectively incorporated demonstrates to an interested party that policies, objectives, and expectations have been met; that emphasis is placed on prevention rather than corrective action; that a company can provide further evidence of reasonable care and regulatory compliance; and that the organization supports continual improvement. The Guidelines recognize that an effective EMS provides an organization with the ability to balance and integrate economic and environmental interests; it describes a number of benefits that may translate into a financial advantage:

- Obtaining insurance at reasonable cost;
- Enhanced image and market share;
- Meeting vendor certification criteria;
- Ability to dispose of waste;
- Cost control;
- Liability limitation and risk reduction;
- Demonstration of due diligence;
- Conservation of input materials and energy;
- Easier site selection and permitting;
- Technology development and transfer;
- Improved industry-government relations;
- Improved environmental performance and state of the environment;
- Meeting customer's environmental expectations;
- Maintaining good public/community relations; community goodwill; and
- Satisfying investor criteria and improved access to capital.

Other reasons include:

- Increased world attention on environmental protection;
- Assurance that an organization is managing its liability in a suitable fashion to protect stockholder interests;
- Certification to EMS standards may become condition of doing business;
- Improvement of an organization's environmental performance through reviews and auditing;
- Increased public pressure for organizations to provide performance information;
- Improved internal management methods; and
- Government requirement of incorporating an EMS.

The concept of proactive rather than reactive is stressed in order to achieve a significant competitive advantage. "This competitive advantage is realized through operational and production efficiencies, loss elimination, and the increased acceptability of products and services in the market. To move from the view that environmental management is just another added cost to the view that it provides for competitive advantage requires new understanding and culture change."[22] Proponents of using the ISO 14000 series as a basis for demonstrating good-faith compliance actions foresee such possibilities as an ISO 14001 certified organization having fewer inspections, fast-track permitting, and mitigation and/or reduction of the severity of penalties for violations.

The ISO 14000 series is not regulatory. In fact, it bars the use of its standards for regulation: "These standards must not be used to...change an organization's legal obligations."[23] With a management rather than a regulatory focus, the standards are not meant to conflict with nor supersede a federal or state environmental law. The ISO 14000 series will not serve as a basis for auditing an organization for compliance with specific environmental regulations but rather provides the parameters in which an organization can develop and expand environmental performance beyond that of baseline regulatory requirements. To be registered under the standard, an organization must have a management system in place to prevent and/or detect violations of environmental laws, regardless of the nation in which the company is operating.

The international trade and competitiveness implications of implementing ISO 14000 may provide a strong impetus for organizations to become certified. ISO 14001 certification may also spawn an organization's growth in overseas operations as well as improved access to international markets. It may be a requirement in order to sell products particularly in Europe.[24] The recently

---

[22] ISO 14000 Guide to Environmental Management Principles, Systems, and Supporting Techniques, 0.2.

[23] *See* 14001, "Introduction. (0.)"

[24] On the surface it would seem that quality is free: Do it right the first time. However, ensuring quality has a price tag. Estimates of costs vary. For example, it has been suggested that for many organizations, preparation for ISO 9000 certification may range up to $500,000 if an organization tracks all its associated costs; for those organizations currently ISO 9000 certified, the cost for ISO 14001 is estimated to be 10% of that figure since the organization has already implemented procedures and documentation of procedures.

signed Uruguay Round of the General Agreement on Trade and Tariffs (GATT) contains provisions encouraging international standards. Implementation of ISO 14001 may also translate into cost-cutting and return on investment: minimization of negative environmental impacts reduce operating costs. Areas such as energy reduction, supply alternatives, waste management, and chemical substitution may provide a return on investment. With daunting numbers of pages of environmental laws and regulations, an EMS provides a mechanism for managing these laws and determining an organization's environmental obligations. Although it is impossible to completely comply with all applicable laws, an effective EMS can greatly increase the compliance rate. Implementation of an EMS demonstrates an organization's concern with environmental issues.[25] It may also show that voluntary compliance can work.[26]

Auditing[27] provides potential cost-savings: auditing of environmental management systems enhance efficiency and, in turn, may provide a competitive edge. Environmental auditing is proactive. It reduces risk. Through environmental audits, problems can be identified internally and corrected quickly. If the environment is harmed, prosecution may look to a pattern of environmental audits as a factor in reducing penalties. Environmental audits provide assurance to management that environmental controls are functioning and provide a focus on environmental issues.

---

[25] The market power of sound environmental principles is seen, for example, in such instances as when Conoco announced it would build two double-hulled oil tankers to help prevent oil spills; this announcement was featured on the front page of the New York Times. And within days of Starkist Tuna's announcement that it would buy only tuna caught by "dolphin friendly" methods, other major tuna canneries followed the policy.

[26] A study by Price Waterhouse LLP, 7 April, 1995, entitled "Voluntary Environmental Audit Survey of US Businesses," found that of 369 respondents, seventy-five percent have an environmental compliance auditing program in place. Ninety-six percent of those that audit do so because "problems can be identified internally and corrected before they are discovered by an agency inspection." Ninety-four percent maintained they audited "to improve our company's overall environmental program and make it proactive."

[27] *See* Price Waterhouse Study, *supra.*

Integration of ISO 9000 and ISO 14000

If an organization currently is certified to an ISO 9000 standard, an obvious question arises: How may the ISO 14000 series be integrated into the existing management system? This is answered by first considering the management system as a whole. The spirit behind the ISO 14000 series is to make certain that all aspects of an organization which may have an impact on the environment, either directly or indirectly, be managed in an environmentally sound way. Very few areas in a management system do not have, in some way, an impact on the environment. Therefore, the need to see the system as a whole rather than as unique and individual blocks of business terrain is the essential first step. To do this, an organization must look carefully at how all sections, all functions, and all divisions of the organization may be effected by environmental issues. To focus on such areas as life-cycle analysis in order to make sounder business decisions in manufacturing products and providing services may involve total re-thinking and re-alignment on the part of management and staff. Environmental impact and environmental effects must be studied and reviewed; this involves a different process than the initial stages of ISO 9000 where no initial "effect of quality" research is needed.

It is imperative that an organization's employees understand the regulatory scheme that affects its operation. With the range of personnel who potentially may be subjected to prison terms for environmental crimes, it is incumbent upon management to develop a system in which there is compliance with all applicable regulations as well as an understanding of these laws. Thus, the first hurdle to overcome is the initial gathering of information with review and analysis.

The ISO 14001 standard is relatively direct. Consequently, once information gathering and analysis is complete, the next step mirrors the techniques of ISO 9000, techniques involved with running any effective management system: policy development, document control, training and educational systems, management review, auditing, continuous improvement, and corrective action. If an organization is already subject to significant environmental regulatory schemes, it may well be that a quasi-environmental management system is already in place.[28] Because auditing to the specifications of each series, ISO 9000 and ISO 14000, is cumbersome and time-consuming, the solution may be a "ubiquitous audit" in which one team of auditors assess all aspects of an integrated management system: quality, safety, and the environment. That solution, of course, requires an auditing team with a background in multiple areas.

Other Environmental Management Standards and Programs[29]

The ISO 14000 Series evolved around the harmonization of standards in order to effect a management strategy that assists in managing existing requirements. Oftentimes, however, other systems develop and include those substantive obligations. The following offers a brief summation of a few of the more salient management schemes.

British Standard 7750 (BS 7750)

The United Kingdom's (U.K.) national standards setting organization, the British Standards Institute (BSI), published an environmental standard, the BS 7750,[30] which was formally launched in 1995; it is the first independently verified environmental management system (EMS) to be put in place. However, organizations have been working with the BSI in BS 7750 pilot programs for several years. The development of BS 7750 began in 1992 and produced a final draft in 1994. Its impetus was derived, in part, from the passage of the U.K. Environmental Protection Act of 1990.

---

[28] For example, the state of Texas' Texas Pollution Prevention Regulation requires a facility to file with the State an executive summary delineating how it will reduce volume and toxicity of waste.

[29] This sampling does not include industry-specific standards such as those developed by the International Electrotechnical Commission (IEC). IEC is working on a guidance document similar to the ISO 14060 Guide. The IEC standard, Guide 02 (C.O.) 594 Environmental Aspects—Inclusion in Electrotechnical Products Standards, contains language that is similar to that found in the ISO 14000 series and, as such, is an international equivalent of ISO for the electrotechnical industry. The ISO 14000 series, however, specifically avoids industry-specific work items.

[30] It is a companion to BS 5750, a standard on quality management systems which was the template for ISO 9000.

The U.K.'s Department of Trade and Industry and Department of Environment has appointed the National Accreditation Council for Certification Bodies (NAACB) as the official accreditor for BS 7750 certification in the U.K.. It has officially accredited ten certification bodies to issue BS 7750 certificates. Both the NAACB and the Dutch Council for Certification (Raad voor de Certificatie [RvC]) are accreditation bodies for BS 7750 in Europe. The standard was developed with the express intention that its requirements should be compatible
with those of the environmental management system specified in the European Community's Eco-Management and Audit (EMA) Regulation. Whether BS 7750 is held to be equivalent to EMA and the European's Eco-Management and Audit Scheme (EMAS) (*see* Section 2, following), remains to be seen if it is seen at all since in 1996, it may be replaced by a European Environmental Management System (EMS), a system incorporating all requirements of EMAS. That standard is being developed by the European standards body, CEN.[31] The core of BS 7750 is its "effects register" requiring organizations to look at the environmental effects of all their activities, decide which are most significant, and draw up objectives and targets for reducing those effects. Progress on these targets is audited regularly. Thus, the effects register provides a concrete strategy for readily identifying and acting upon environmental problems. It also allows an auditor to determine if an organization has a sound method for determining what it considers significant effects.

In addition to compliance with all relevant legislative and regulatory requirements, other objectives and targets must be identified after consideration of the environmental effects register, the financial, operational, and business requirements of the organization, and the views of relevant interested parties. The objectives and targets must be consistent with the environmental policy and must quantify wherever practicable the commitment to continual improvement in environmental performance over defined time-scales. BS 7750 requires formalized training and stringent communication. Section 4.3.4—personnel, communication and training—requires that employees or members

[31] The CEN is looking not only to BS 7750 but is focusing on EMS standards in France, Ireland, and Spain as well as the ISO 14000 series. If the ISO 14000 series meets the requirements of EMAS, CEN may adopt it as the European standard. If the ISO 14000 series does not fill the bill, then CEN may use it as a base; it could then create a separate package of additional standards to pull the ISO 14000 series in line with EMAS.

of a facility at all levels are to be aware of the environmental policy and objectives, significant environmental effects of their work activities, their roles and responsibilities in achieving compliance, and the potential consequences of deviation from standard operating procedures.[32]

How does BS 7750 compare with the ISO 14000 series? Many of the ISO 14001 standards' provisions are based on BS 7750. And like ISO 14001, it reviews progress by performing an audit of the organization's system. BS 7750 does not make reference to legislation in its specification standard. With both standards, third-party certification is optional; and certification details are not specified. However, BS 7750 is less flexible and more prescriptive than ISO 14001. As such, it may not be appropriate for global application. Smaller businesses in lesser developed countries or newly industrialized countries may not have the resources to implement this standard. The ISO 14000 series has nothing to compare with the BS 7750 effects register. BS 7750 mandates public disclosure of corporate environmental plans and objectives. This could equate into a disincentive to establish hard-hitting, far-reaching environmental goals. Another problem could arise if it becomes an "industry standard": The U.S. Sentencing Guidelines refer to "industry standards" as a factor in assessing environmental penalties. With its inflexibility and complexity, BS 7750 may not be the appropriate bench-mark standard.

Eco-Management and Audit Regulation (EMAR) and the Eco-Management and Audit Scheme (EMAS)[33]

The Eco-Management and Audit Regulation (EMAR) contains the Eco-Management and Audit Scheme (EMAS); it is a European Union (EU)[34]

---

[32] For a case study review of why a Canadian company chose BS 7750, *see International Environmental Systems Update,* vol. 2, no. 5.

[33] The European Union set requirements with its Eco-Management and Audit Regulation (EMAR). EMAR is a regulation that all Economic Union member nations have agreed to adopt and is the written language of the regulation while EMAS is a systems approach to the regulation; the terms have been used interchangeably, although technically, EMAS is the part of EMAR that refers to EMS elements.

[34] The EU is composed of fifteen European member nations. The EU "legislative bodies" are: The European Commission; the European Council; the European Parliament; the Economic and Social Committee; and the Committee of the Regions. Its documents include: regulations, directives, decisions, and recommendations.

standard which came into force in 1995 and includes a rather elaborate management and auditing scheme as well as detailed performance standards, third-party verification, and public statements which, according to EMAS, "follow the initial environmental review and the completion of each subsequent audit or audit cycle." It seeks to help those in the manufacturing industry through promoting environmental management and informing the public about the performance of companies. It was approved by the environmental ministers of the EU in 1993.

EMAS is site-specific and, at this point, its adoption is voluntary. To achieve registration, eight steps must be completed:

- Develop a corporate environmental policy;
- Carry out an environmental review of the site;
- Introduce an action program and an environmental management system;
- Carry out regular environmental audits to evaluate how the system is working;
- Set responsive objectives based on the results of the audit;
- Prepare an environmental statement concerning the audited site;
- Submit all the above data for independent verification; and
- Send the above to its national authority.

As with BS 7750, EMAS has an effects register in which companies must clearly identify the environmental effects of all their activities, decide which are most significant, and draw up objectives and targets for reducing those effects. Progress on these activities is made available to the public.

Benefits derived from EMAS registration may include reduced insurance premiums, fewer local environmental inspections, and financial savings through energy efficiency or waste minimization. At the same time, however, companies must consider the worth of submitting to the public an environmental statement as well as the value of both increased costs in meeting the requirements and receiving outside verification.

There have been problems with some member states establishing competent bodies to implement the program and with the accreditation body to carry out the certification. EMAS is a regulation and, as such, does not require member state legislative action for implementation. The problem arises, however, when legislation is required in a member state to establish an accreditation body. Also, there are discrepancies among member states as to who may verify: in some member states only individuals can be verifiers; in others, individuals or companies may assume this role.

How does EMAS compare with the ISO 14000 series? Both systems require top-level management commitment, environmental policy, procedures, training and auditing. EMAS only applies across the EU while ISO 14001 is international in scope. EMAS is site specific while ISO 14001 has a stronger focus on products and services, operating procedures, and can apply to organizations as well as specific sites; ISO 14001 is less limited in its scope of activities.[35] EMAS is more prescriptive in some areas and defines environmental performance in terms of best available technology. It also has a stronger emphasis on public communication. The Draft International Standard (DIS) of ISO 14001 reflects the addition of pollution prevention and compliance auditing language which brings it more in line with the requirements of the EMAS.[36]

An area of concern is whether the ISO 14001 standard will be recognized by EMAS: EMAS is centered around *performance* while ISO 14001 centers on *conformance.* Under ISO 14001, an auditor looks to see if the system conforms to the criteria and not whether a system is adequate. Unlike ISO 14001, EMAS requires that a verifier, someone independent of the organization, validate an organization's claim that it is meeting the guidelines of EMAS and that the statement is reliable. Third-party verification is mandatory with such duties detailed. Frequency of audits is specified in EMAS but not in ISO 14001. Additionally, the same problems identified with the BS 7750 system may arise. (*See* BS 7750 discussion in Section 1, this chapter.)

Responsible Care (RC)

RC is the chemical industry's voluntary health, safety and environmental program. It was launched by the Chemical Manufacturers Association (CMA) in 1988 and participation by individual businesses is an obligation of membership in CMA. RC lists management practices that chemical companies should adopt in such areas as emergency response, pollution prevention,

[35] Theoretically, EMAS should not affect a U.S. company that exports to Europe but has no sites in Europe.

[36] The ISO 14001 prevention of pollution language was added to section 4.2.3, Objectives and Targets; the compliance language was added to section 4.4.1., Monitoring and Measuring.

employee risk, and health and safety awareness. Over one-hundred specific actions are detailed by RC. Its focus is on performance and has the following goals:

- Improved chemical process;
- Enhanced practices and procedures;
- Reduction of waste, accident, incident, and emission;
- Reliable communication and dialogue; and
- Heightened public involvement and input.

Its six Codes of Management Practices include:

- Product stewardship;
- Community awareness and emergency response;
- Process safety;
- Employee health and safety;
- Pollution prevention; and
- Distribution.

How does ISO 14001 compare with Responsible Care? Although ISO 14001 and RC are similar in their goals, ISO 14001 and RC differ in several aspects, the most notable of which is *performance* (RC) versus *conformance* (ISO 14001). While there is overlap between ISO 14001 and RC's pollution prevention code, ISO 14001, on the whole, is much broader than RC: an organization using RC alone will not meet ISO 14001 requirements without additional work. While RC includes numerous specific management actions, the ISO 14001 may be used as a system to manage these actions. As well, ISO 14001 addresses the management of items not contained in RC. The ISO 14001 standard requires identification of all aspects of an operation affecting the environment; RC focuses on the impact of chemicals. RC also has no specific monitoring and measurement standards as does ISO 14001. RC does not contain a management system of EMS audits and review of audits by senior management as does ISO 14001. ISO 14001 may be a way to register or certify that an organization is adhering to the principles of RC and managing its specified actions; but it must be determined whether ISO 14001 brings the kind of third-party assessment that might lead to increased credibility for RC.

Because RC is entrenched in the chemical manufacturing business structure, and because ISO 14001 may not be the vehicle to verify RC systems, chemical companies may not feel justified in seeking registration to ISO 14001. Further, to separate the two may translate into a series of audits, each assessing to a different standard but covering the same territory.

Environmental Protection Agency's Environmental Leadership Program (ELP)

This year-long program is designed to encourage regulated entities to develop innovative auditing, disclosure, and compliance methods. The EPA intends to use what it learns to streamline federal and state reporting requirements and reduce inspections. Pilot projects were created with companies and public agencies to test criteria for auditing and certification of voluntary compliance programs. "If these pilot projects are successful, standards developed through ELP could lead to reduced inspections and public recognition for companies or agencies with state-of-the-art compliance programs."[37]

The ELP is a voluntary program in which participants develop their own EMS which meet EPA guidelines. The incentive is that the EPA may offer the regulated entities leniency in areas of non-compliance. Under the ELP, the EPA will not perform routine enforcement inspections at the participant's facility. In exchange, the facility must take steps to ensure they are complying with environmental standards. Facilities must have agreed to disclose to the EPA and the States any actual or suspected violations that arise. The Agency will then give participants ninety days to correct their non-compliance without filing any enforcement action against them. Participants also have agreed to give the EPA the results of any relevant environmental compliance audits conducted within the past two years. The EPA may conduct inspections if it receives a complaint about a possible violation. The results may be used in Project XL—Excellence in Leadership—a new EPA project which is part of President Clinton's regulatory reform package for the EPA.[38]

---

[37] *Federal Register,* 3 April 1995.

[38] Under Project XL's pilot program, some businesses will be released from certain regulatory requirements if they exceed current environmental standards.

The ISO 14000 series could supply the systematic approach to compliance required by the ELP. Further, participants in the program may well be on their way to achieving certification after having implemented ELP's requirements. The twelve participants include: Arizona Public Service, Deer Valley Facility, Phoenix, Arizona; Ciba-Geigy Corporation, St. Gabriel, Louisiana; Duke Power Riverbend Steam Station, Mt. Holly, North Carolina; the Gillette Company (in three manufacturing centers); the John Roberts Company, Minneapolis, Minnesota; McClellan Air Force Base, Sacramento, California; Motorola Incorporated, Oak Hill Facility, Austin Texas; Ocean State Power, Burrillville, Rhode Island; Puget Sound Naval Shipyard, Bremerton, Washington; Salfe River Project, Phoenix Arizona; Simpson Tacoma Kraft Company, Tacoma, Washington; and WMX Technologies Incorporation, Arlington, Oregon.

Environmental Protection Agency's Common Sense Initiative (formerly known as the Green Program)

This Initiative was first announced in November, 1992. It takes a holistic approach to enforcement and compliance and involves a roundtable format for building consensus and reforming the "command-and-control" system.[39] Its aim is to improve the environmental management system by customizing environmental regulations across all media for a particular industry sector. According to the EPA, it is a "first look into the future of environmental protection," bringing all the affected parties to the table to review and improve the current system. Six industries were selected for the pilot project: auto manufacturing, printing, computers and electronics, petroleum refining, metal finishing, and iron and steel. The subcommittee and workgroups also include representatives from federal and state government, environmental groups, and labor organizations. These teams were to review new regulations and make "consensus recommendations" to make it easier for industry to comply through simplification of reporting while, at the same time, increasing pollution prevention in such ways as promoting environmental technology innovation.

With its emphasis on all media, the ISO 14000 series may be an appropriate tool to be used in developing the holistic approach to permitting

[39] Canada uses as similar system.

under the Initiative. It may also provide a basis for companies to systematically approach pollution prevention.

American Society for Testing and Materials (ASTM) Environmental Management System

ASTM has developed two compliance auditing standards to assist organizations in understanding whether they are in compliance with U.S. legal requirements. They are intended to reflect U.S. industry practices. The first is Practice for Environmental Compliance Audits (PS11). The second is Guide for the Study and Evaluation of an Organization's Environmental Management System (PS12).

The objectives of PS 11 are to document minimum requirements for compliance audits, both internal and external, identify legal issues associated with such audits, and provide minimum criteria for auditor qualifications. PS 11 provides a description of accepted procedures associated with environmental regulatory compliance audits. Its appendix includes legal citations and information on certain regulatory bodies.

PS 12 defines acceptable guidelines for the study and evaluation of environmental management systems. Like PS 11, it is based on U.S. regulations and programs such as the EPA's Environmental Leadership Program, as well as state programs.

How does this system compare with the ISO 14000 series? PS 12 is concerned with identifying whether an industry has an EMS in place. EMS is defined differently than in the ISO 14000 series: Environmental Management Systems are "those policies, procedures and activities effected by the entity's board of directors, management and other personnel, which are designed to provide reasonable assurance that environmental objectives are being achieved."[40] The ISO 14000 series defines EMS as "that part of the overall management system which includes organizational structure, planning activities, responsibilities, practices, products, processes and resources for developing, implementing, achieving, reviewing, and maintaining the environmental policy."[41]

---

40 PS 12, Clause 3.2.1.

41 ISO 14001, 3.6.

Unlike PS 11 and 12, the ISO 14000 series is directed toward international application rather than being country-specific. ISO 14001 is used to develop and maintain an EMS with auditing focused on the system itself rather than on regulatory compliance auditing. Other ASTM environmental performance standards under development include: Practice for Voluntary Cleanups at Contaminated Sites; Guide for Environmental Laboratory Results Export Format; Guidelines for the Conduct of Corporate Fiduciaries in Environmental Matters; Guide for Evaluating the Environmental Health of a Residence; Guide or Practice for Environmental Audits for Residential Real Estate; and Guide for Remediation Project Management.

International Safety Management Code (ISM)

Because approximately eighty-five percent of marine casualties may be attributed to human error, a management strategy for the safe operation of ships is essential. The International Management Code for the Safe Operation of Ships and for Pollution Prevention, which is also called the "International Safety Management (ISM) Code," is a document addressing shipboard and shore-based management for the safe operation of ships and for the prevention of marine pollution. The ISM Code focuses on a management system specifically related to safety and pollution prevention with responsibility assigned to the owner or whoever has assumed responsibility for operation of the vessel from the owner. The ISM Code adopted by IMO as Resolution A.741(18) on 4 November 1993 "strongly urges governments to implement the ISM Code on a national basis, giving priority to passenger ships, tankers, gas carriers, bulk carriers and mobile offshore units, which are flying their flags, as soon as possible but not later than June 1, 1998, pending mandatory applications of the code." At present, U.S. certification is voluntary. However, the ISM Code has been adopted into the International Convention for the Safety of Life at Sea (SOLAS 74),[42] and will be mandatory for certain vessels beginning 1 June 1998.

---

[42] SOLAS Chapter IX: Management for the Safe Operation of Ships. The Conference adopted this new Chapter with revised definitions of ship types and application dates (1 July 1998 for all passenger ships, and the following vessel types of 500 gross tonnage and over: oil tankers, chemical tankers, gas carriers, bulk carriers, and cargo high speed craft; and 1 July 2002 for all other cargo ships and Mobile Offshore Drilling Unit's of 500 gross tonnage and over). It should be noted that the new requirements apply only to self-propelled MODU's.

The ISM Code is intended as a means to encourage continuous improvement of safety management skills in the maritime industry and can be applied to all ships. It addresses the importance of designated persons with various responsibilities within the company and requires that management procedures be documented, coordinated, and monitored in accordance with government and company requirements. The result is the company's Safety Management System.

"The ISM Code is intended to re-orient the current approach to regulatory compliance from the industry's passive defect notification and correction response mode to an aggressive approach to safety. Under a proactive approach, potential discrepancies are resolved by the companies themselves, before they become significant safety or environmental problems."[43] The ISM Code is expressed in broad terms so that it can have widespread application. Different levels of management, whether shore-based or at sea, will require varying levels of knowledge and awareness of the ISM Code content.

As stated in the preamble: "The cornerstone of good safety management is commitment from the top. In matters of safety and pollution prevention it is the commitment, competence, attitudes and motivation of individuals at all levels that determines the end result." The ISM Code requires management and responsible personnel to be accountable which, in turn, ensures that their vessels are operated in conformity with applicable standards. A Document of Compliance (DOC) is issued to a qualifying company when adherence to ISM Code provisions has been confirmed by a third party. Vessels owned or operated by a company which has a DOC may be issued a Safety Management Certificate (SMC) if they meet applicable ISM Code provisions. Some companies already have management practices instituted which reflect elements of the ISM Code: a company's personnel manual, emergency response plans, and training programs may already be consistent with ISM Code specifications; compliance for such companies may require minimal effort. At the other end of the spectrum, a vessel traveling to international ports must be cognizant and in compliance with the individual nation's regulations and maintain appropriate certificates and documentation on board.

[43] The United States Coast Guard Navigation and Vessel Inspection Circular No. 2-94, 15 March 1994.

Until such compliance dates specified in SOLAS Chapter IX are triggered, certification is voluntary. In the U.S., the American Bureau of Shipping (ABS) has developed procedures to evaluate shipping operations against the requirements of the ISM Code. At this time, the ABS may issue appropriate documentation to qualifying companies and vessels.[44]

International Ship Management Association (ISMA) Code

The ISMA was established in April, 1991, in response to the need to achieve and maintain the highest standards of ship management practice. Currently, there are over thirty members. Three classification societies with plans to expand to six are responsible for assessing and verifying quality-assured management of ISMA members. An integral part of the formation of ISMA was the adoption of the Code of Ship Management Standards through which quality and safety can be enhanced. Although the Code was originally intended for independent ship management companies, it may be adopted by shipowners operating their own vessels. Unlike the regulatory requirements for International Safety Management Code which involves all vessels, the ISMA Code is voluntary and applies to enterprises engaged in the business of ship management. From an operational perspective, however, the ISMA Code is far-reaching, having a potential impact on all areas of ship management.

The Code specifies requirements for quality assured ship management and operation. It identifies those areas where systems and controls are essential to meet its objectives and the requirements apply to both shore-based and shipboard management. An important objective is to provide quality assured ship management services which includes: "operating the ship and transporting cargo safety and efficiently; and conserving and protecting the environment." Provisions for updates and revisions are provided in the Code: classification societies and the ISMA Code Committee meets at least once a year to participate in this process.

[44] The United States Coast Guard Navigation and Vessel Inspection Circular No. 2-94, 15 March 1994.

3
ISO 14001: Environmental Management System Specification

There ain't no rules around here! We're trying to accomplish something!
Thomas Edison (1847-1931)

Introduction

Today, Tom might well find himself in jail holding his light bulb if his accomplishments seriously degraded the environment. Had he set up his business using the ISO 14001 specification document, he may have avoided an environmental problem.

The ISO 14000 series may be divided into two groups: guidance documents and specification documents. Guidance documents include words such as *should* while the specification document typically includes words such as *shall* or *must*. When a standard uses the word *shall* or *must*, then the standard is expressing a requirement. For example, in ISO 14001, Section 4.1., Environmental Policy, the specification document details what the environmental policy *shall* contain. In other words, all of the specific items enumerated must be contained within the organization's environmental policy. ISO 14004 is a guidance document; as such, it is written in language peppered with recommendations: "an organization *should*...."

ISO 14001 is the specification document and, as such, is the standard against which an organization's environmental management system (EMS) will be judged; organizations may become certified against this standard. It is not a technical product standard: instead, the management system in place for the production of a product may be certified. An organization achieving certification may proclaim that it has implemented a documented environmental management system that is consistently followed. Certification does not imply an organization produces more environmentally sound products.

The core elements of the EMS include:

- Policy;
- Planning;
  - Environmental Aspects
  - Legal and Other Requirements
  - Objectives and Targets
    - Environmental Management Program
- Implementation and Operation;
  - Structure and Responsibility
  - Training, Awareness and Competence
  - Communication
  - Environmental Management System Documentation
  - Document Control
  - Operational Control
  - Emergency Preparedness and Response
- Checking and Corrective Actions; and
  - Monitoring and Measurement
  - Non-conformance and Corrective and Preventative Action
  - Records
  - Environmental Management System Audit
- Management Review

To become ISO 14001 certified, an organization must develop an environmental policy; establish procedures to identify environmental aspects of all activities in order to determine which have significant impact; document environmental objectives and targets; include a commitment to the prevention of pollution; train all personnel whose work may significantly affect the environment; create an auditing system to ensure the program is properly maintained; and review the audit results to ensure continual improvement. The standard does not require organizations to meet top-level performance standards but rather to standardize the field of environmental management tools and systems. The following are its core principles:

- Commitment to environmental management, one of the highest corporate priorities, should permeate the organization through all levels;
- Identification of legislative and regulatory requirements and environmental aspects associated with the organization's activities, products, and services;
- Development of a management process for achieving objectives and targets;
- Providing appropriate financial and human resources to achieve objectives and targets;
- Assignment of clear procedures for accountability and responsibility;
- Establishment of a management review and audit process to identify areas for improvement; and
- Development and maintenance of communication with internal and external interested parties.

For certification or self-declared conformance to ISO 14001, an organization must comply with all requirements of ISO 14001. This standard specifies the requisites of an environmental management system and is divided into the following sections:

Introduction (0.)
Scope (1.)
References (2.)
Definitions (3.)
Environmental Management System (4.)
Annexes

An informative Annex A provides guidance on the use of the specification. Annex B contains references.

Prefatory Sections

The prefatory sections provide the general orientation to ISO 14001 such as that the specification document contains only those requirements that may be objectively audited for certification or self-declaration purposes. The environmental performance criteria are twofold: criteria are comitant to comply with all applicable law and comitant to continually improve.

The core requirements for environmental management systems are specified. As well, an organization must comply with regulatory requirements and must consider significant environmental impacts. The standard is appropriate for those organizations that wish to implement, maintain and improve its EMS; assure itself of conformance; demonstrate conformance to others; seek certification or registration from a third-party and/or self-declare conformance with the standard.

The definition section establishes definitions for the purpose of this standard and mirrors the definitions found is ISO 14004.

The Core Sections: Developing and Implementing the EMS

An organization's ability to conform to the ISO 14001 standards is assessed through the Environmental Management System which consists of five main paragraphs, each with subsections. In outline form, what is to be emphasized in an EMS is visible:

- Environmental Policy (4.1.)
- Planning (4.2)
    - Environmental aspects (4.2.1.)
    - Legal and other requirements (4.2.2.)
    - Objectives and targets (4.2.3.)
    - Environmental management programs (4.2.4.)
- Implementation and Operation (4.3.)
    - Structure and responsibility (4.3.1.)
    - Training, awareness, and competence (4.3.2.)
    - Communication (4.3.3.)
    - Environmental management system documentation (4.3.4.)
    - Document control (4.3.5.)
    - Operational control (4.3.6.)
    - Emergency preparedness and response (4.3.7.)
- Checking and Corrective Actions (4.4.)
    - Monitoring and measurement (4.4.1.)
    - Non-conformance and corrective and preventive action (4.4.2.)
    - Records (4.4.3.)
    - Environmental management systems audit (4.4.4.)
- Management Review (4.5.)

These sections will be highlighted with comments incorporating the standard itself, an annex ("The Guidance for Use of the Specification, Annex A"), as well as practical notes.

Environmental Policy

The environmental policy section (4.1.) may be the most fundamentally important section of ISO 14001 in terms of the framework for implementing the standard. Development of a policy is the step which galvanizes the organization and focuses on addressing environmental issues. It is the source from which flows other elements: objectives, targets, implementation, audits, review, and continuous improvement. How these other elements reflect the policy is one of the questions asked in an audit.

Environmental policy is defined as the "statement by the organization of its intentions and principles in relation to its overall environmental performance which provides a framework for action and for the setting of its environmental objectives and targets."[45] Details of what that environmental policy shall contain is given. Top management must ensure that the organization's environmental policy is appropriate and is designed as the basis on which environmental objectives and targets are set. As with ISO 9000, an organization's environmental policy must be documented, maintained and communicated to employees. Interested external parties must have access tot this policy. Although a verbatim inclusion of the standard's wording is not necessary in the policy statement, the underlying issues must be included.

Senior management may begin policy development by looking at the mandates and deciding what each means. For example, when senior-level managers establish the organization's environmental policy, it must be appropriate to the environmental impacts of its activities. Senior management would look at its organization as a whole and at each of its parts to discern how the environment may be affected by what it does and what it produces. This would include thinking about the actual operational sites and the impact, both directly and indirectly, that a product or service may have on the environment. The standard requires continual improvement. In thinking "down the line," continual improvement may be evidenced by implementing sounder and/or more sensitive environmental procedures.

---

[45] Section 3.10, ISO 14001.

A first draft may articulate a commitment to fulfill each mandate. From this, an organization may focus on potential objectives and targets of the policy to see how the policy may be implemented; in tandem, an organization should be mindful of EMS implementation across-the-board. Such a review will keep the policy from being too broad and/or too vague and may eliminate future difficulties with developing objectives, targets, and other aspects of the EMS to fit the policy. Avoiding broad statements is also important because such statements leave room for subjective judgments by the auditors. Organization-specific items can be included with the same review of implementation of future objectives and targets. Consideration should be given to ideas such as recycling; purchasing environmental friendly supplies; and providing clients or customers with suggestions on environmental sound use and/or disposal of the product. The policy should avoid detailed requirements; it should be a document from which managers may set priorities. A policy statement should be succinct rather than verbose, straight-forward rather than complicated.

The standard is clear regarding the concept of regulatory compliance: An organization must be dedicated to operating in compliance with all applicable laws in all aspects of its business. For example, a policy statement may read: *Organization X will meet and exceed legislative requirements.*

The standard is not as clear when providing the framework for setting and reviewing environmental objectives and targets; the statement could be couched in terms of: *Organization X sets objectives which comply with all applicable regulatory requirements and will be supported by internal targets.* That the policy must be documented, implemented, and communicated to all parties is certain. The policy must also establish methods that communicate or make available its environmental policy to those outside the organization: neighbors, investors, shareholders or simply those making a research inquiry. In other words, the policy must be reasonably accessible.

Planning

The planning section (4.2.) is divided into four subsections:

Environmental aspects (4.2.1.)
Legal and other requirements (4.2.2.)
Objectives and targets (4.2.3.)
Environmental management programs (4.2.4.)

The section provides a process by which organizations can identify significant environmental aspects that should be addressed as priorities. What constitutes "environmental aspects" has led to some confusion. Section 3.3 defines them as "elements of an organization's activities, products, and services which can interact with the environment." An organization must implement and maintain a systematic approach to identify potential environmental aspects—benefits as well as negative impacts—in all phases of its business. Thus, in defining the parameters for environmental aspects, an organization includes the environmental impact of its activities as well as the materials it uses and produces. These must be prioritized. Areas that traditionally were not geared toward assessing environmental impacts such as marketing, profits/financial, and site-specific areas must now identify any such potential impacts. New and current operations as well as new and current materials must be screened for environmental soundness. For example, procurement has an added environmental dimension.

According to the corresponding Annex section, for organizations with no existing EMS, the first step should be to establish its environmental position by means of a review in which environmental aspects are considered for all operating conditions: normal, shut-down, start-up, and emergencies. Such a review should cover legal requirements; resulting information gleaned from previous incidents; and present environmental management strategy and procedures. The reviews may be in the form of checklists or interviews as well as on-site assessment.

For many organizations, the development of a database may prove a valuable management tool for prioritizing and assessing large quantities of information and assisting in environmental reviews. This is not a life-cycle review: an organization need not evaluate each product, component or raw material input. It may select categories of services, activities, or products to identify those aspects most likely to have a significant impact. Depending on the level of control that an organization has over a product or potential product or service, the organization should strive to consider the environmental aspects such as contamination of air, water and land; impact on sensitive areas such as wetlands; and the use of natural resources.

The legal section (4.2.2.) is unambiguous: An organization must maintain a procedure that will identify and maintain access to all legal and other requirements such as industry codes of practices; agreements with public authorities; and non-regulatory guidelines which apply to its activities, products, and services. The key words are "identify" and "have access to requirements."

Whereas the policy section is clear as to what must be addressed in the policy, the objectives and targets section (4.2.3.) is less certain. Reference to the ISO 14004 Guidance document, however, does provide more direction (see Chapter Four). In general, organizations may support concepts such as reducing pollution; however, this will not satisfy the requirements of the standard. More specific objectives such as *reducing the release of CO_2 emissions 10% by year X* are required. Objectives and targets must be documented and maintained for each relevant function and level; and they must be consistent with the environmental policy including the commitment to the prevention of pollution and comply with al legal requirements. The standard notes that the practical aspects of an organization's business—technological, financial, and operational—be considered in formulating attainable objectives. Once the objectives and targets are identified, the means and time schedule for attaining them must be incorporated into the environmental management program.

If a project relates to new development or new or modified activities, products, or services, the program must be amended where relevant to ensure that environmental management applies to such projects.

Implementation and Operation

The implementation and operation section (4.3.) contains seven subsections:

Structure and Responsibility (4.3.1.)
Training, awareness, and competence (4.3.2.)
Communication (4.3.3.)
Environmental management system documentation (4.3.4.)
Document control (4.3.5.)
Operational control (4.3.6.)
Emergency preparedness and response (4.3.7.)

Identification of who's responsible for what is required as well as communication and documentation of these roles and responsibilities at each level and/or function; and resources must be available to support an effective EMS including human as well as financial and technical resources. In particular, top management must appoint a representative responsible for making certain the EMS requirements are maintained for accountability to top management through review.

Training needs must be identified and all personnel whose work may have a significant impact on the environment must be competent on the basis of education and appropriate training and/or experience. Procedures must be maintained to make employees aware of their responsibility in ensuring that the EMS requirements are fulfilled. They should also be made aware of the importance of conforming with the EMS policy and procedures and become cognizant of the potential consequences of non-compliance with these requirements.

Procedures must be maintained for both internal and external communications regarding the EMS. Internally, this includes communication among various functions and levels of the organization. They must be maintained for receiving, documenting, and responding to relevant communication from external interested parties. Processes for external communication on significant environmental aspects must be considered and that consideration recorded. A dialogue with interested parties concerning issues such as emergency response may be appropriate.

Information, paper or electronic, describing the core elements of the EMS and their interaction and providing direction to related documentation must be maintained. A single manual is not required. Related documentation could include items such as organizational charts; operational procedures; and site emergency plans.

A procedure for the control maintenance of all documents required by this standard must be established to ensure that documents are locatable, reviewed and revised or removed as needed, and available where appropriate. These documents must be legible, dated, readily identifiable, and maintained in an orderly manner.

Identification must be made of those operations and activities that are associated with significant environmental aspects in line with the policy, objectives and targets. Unfortunately, no Annex guidelines are given, but the intention is that once identified, these activities can be targeted and documented procedures in place to cover situations where their absence could lead to deviations from the environmental policy and the objectives and targets. Further, operating criteria must be described in the procedures. Procedures related to the significant environmental aspects of goods and services must be maintained; and relevant procedures and requirements must be communicated to suppliers and contractors.

Maintenance of procedures identifying the potential for and the response to accidents and emergencies as well as preventing and mitigating environmental impacts is required. There must be review and revision of these procedures following an incident; as where practicable there must be periodic testing of these procedures.

Checking and Corrective Action

The checking and corrective action section (4.4.) contains four subsections:

Monitoring and measurement (4.4.1.)
Non-conformance and corrective and preventive action (4.4.2.)
Records (4.4.3.)
Environmental management system audit (4.4.4.)

A procedure must be maintained in which all key elements of operations and activities that can have a significant impact on the environment are monitored and measured. This includes recording information to track performance, relevant operational controls, and conformance with objectives and targets.

Monitoring equipment must be calibrated and maintained. Records of this process must be retained according to procedure. As well, procedures to periodically evaluate compliance with relevant environmental legislation and regulations must be maintained. However, it should be noted that there is no mention of documentation in this latter area.

Initiation of corrective and preventive action is required when auditors find non-conformance evidence that the standard's requirements are not being met as well as when tracking and completing preventive action. Procedures must be established not only to prevent non-conformance, but to investigate and mitigate any non-conformance. Any changes in the documented procedures resulting from corrective and preventive action must be implemented and recorded.

Procedures for the safe-keeping, retention, and disposition of environmental records including records of training, audits, and reviews must be maintained. As with documents in general, such records must be legible, identifiable, readily retrievable, and traceable to the area or product involved.

An audit of the specification document by impartial external or internal persons is one of the key elements of continual improvement. Periodic audits of the EMS are required to assess whether the EMS conforms to EMS procedures and comports with the requirements of the standard. The result of such audit must be given to management. The audit program must be based on the environmental importance of the activity as well as the results of previous audits. Control of the audit includes determining the "who, what, when, where and how" of the audit process as well as the reporting requirements.

Management Review

Senior managers must review the EMS for suitability, adequacy, and effectiveness at appropriate intervals to ensure continual improvement. Because an organization is a dynamic entity, changes in the EMS policy objectives, or other elements any be necessary. Thus, an EMS must be adaptable and reflect the current status of an organization's business strategy. The basis for the review may include such factors as audit results or business strategy or condition changes. The management review must be documented.

The Annexes

The Annex A guideline, "Guidance on the Use of the Specification," provides clarification and is not intended for use in audits. It is divided into the following sections:

- Environmental Management System (A.4.)
- General (A.4.0.)
- Environmental policy (A.4.1.)
- Environmental Planning (A.4.2.)
  - Environmental Aspects (A.4.2.1.)
  - Legal and other requirements (A.4.2.2.)
  - Objectives and targets (A.4.2.3.)
  - Environmental management program (A.4.2.4.)
- Implementation and Operation (A.4.3.)
  - Structure and responsibility (A.4.3.1.)
  - Training, awareness and competence (A.4.3.2.)
  - Communication (A.4.3.3.)
  - Environmental management system document (A.4.3.4.)
  - Document control (A.4.3.5.)
  - Emergency preparedness and response (A.4.3.7.)

Checking and Corrective Action (A.4.4).
- Monitoring and measurement (A.4.4.1.)
- Non-conformance and corrective and preventative action (A.4.4.2.)
- Records (A.4.4.3.)
- Environmental management system audit (A.4.4.4.)

Management review (A.4.5.)

Annex B gives references to ISO 9000 and ISO 14004.

4
ISO 14004: Guidelines for Environmental Management Principles and Supporting Techniques

If a better system is thine, impart it; if not, make use of mine.
Horace

Background

ISO 14004 is a guidance document, a document written in the language of advice that *may* be taken rather than advice that *must* be taken. ISO 14004 explains environmental management concepts, defines key terms, and provides practical advice for the design and implementation of an environmental management system (EMS). It is more detailed and has a broader scope than ISO 14001, the specification document. This chapter takes the reader through the most important sections of the Guidelines. Although it follows the general arrangement of the document, it is interspersed with discussion about each topic. Interpretive comments have been kept to a minimum.

The heart of this document is its five principles. Each of these correspond to a section in the specification standard, ISO 14001, and is noted in parenthesis:

Principle 1: Define its environmental policy and ensure commitment to its EMS (Section 4.1. Environmental Policy).

Principle 2: An organization should formulate a plan to fulfill its environmental policy (Section 4.2. Planning).

Principle 3: For effective implementation, an organization should develop the capabilities and support mechanisms necessary to achieve its environmental policy objectives and targets (Section 4.3. Implementation and Operation).

Principle 4: An organization should measure, monitor, and evaluate its environmental performance (Section 4.4. Checking and Corrective Action).

Principle 5: An organization should review and continually improve its environmental management system with the objective of improving its overall environmental performance (Section 4.5. Management Review).

Various sections offer issues to be considered in achieving compliance with that section or subsection.

ISO 14004 is divided into:

Introduction (0.)
Scope (1.)
Normative References (2.)
Definitions (3.)
Environmental Management System (EMS) Principles and Elements (4.)
    How to Start (4.1.)
    Planning (4.2.)
    Implementation (4.3.)
    Measurement and Evaluation (4.4.)
    Review and Improvement (4.5.)
Appendix

The Guidelines are intended to be consistent with the goal of sustainable development and are compatible with diverse cultural, social, and organizational frameworks.

Description

The more significant sections of the guidance document are highlighted herein.

Overview

Environmental reviews and audits assess whether an organization's environmental performance meets the requirements of legislation and its own policies and need to be conducted within a structured management system integrated with overall management activity that address relevant characteristics of the desired environmental performance. The Overview stresses that environmental management is an integral part of an organization's overall management responsibility and, as such, should be coordinated with existing efforts in other areas. The EMS should be a living document subject to on-going, interactive processes consisting of defining, documenting, and continually improving. ISO 14004 emphasizes the importance of small and medium-sized enterprises (SMEs), recognizing that such enterprises may have unique needs.

The Overview suggests ten key criteria for managers who are implementing an EMS. Although they do not purport to be all-inclusive, they do provide a wide canvas on which to draw a fundamental environmental policy. Among others, the focus is on an EMS being one of the highest priorities of an organization; the importance of dialogue with interested parties; determination of legal requirements; identification of an organization's environmental aspects; accountability; strategic planning; appropriate resources; and review and audits.

The ISO 14004 may be incorporated as a vehicle to achieve third-party certification, self-declaration, or contracting party recognition. The later may be appropriate for small businesses with a familiar client base. The document may be used to simply improve an EMS or develop one.

Related documents such as 14011—Auditing of Environmental Management Systems—may be utilized in the overall implementation of the EMS. Factors influencing such a decision are varied and may depend on the size, marketing position, environmental strategy, degree of sophistication of an existing EMS, and other factors.

Benefits of an EMS

An organization's effective EMS may translate into providing assurance to interested parties that environmental policies and environmental expectations are met; that a pro-active stance is the norm; and compliance with laws coupled with continuous improvement are among the organization's objectives that have been met. Economic benefits[46] may include easier access to investment capital and cost reduction for insurance. The general goodwill of customers and the community who recognize the commitment of an organization to an effective EMS and to solving environmental problems may result in increased market share for its services and/or products. Of course, lessening of exposure to liability ultimately effects the bottom-line.

"How To": Environmental Management System's Principles and Elements

Section 4 is the heart of the Guidelines, the "How To." It is divided into five Sections:

How to Start: Commitment and Policy (4.1.)
Planning (4.2.)
Implementation (4.3.)
Measurement and Evaluation (4.4.)
Review and Improvement (4.5.)

How to Start: Commitment and Policy

Like the other subsections, this one is broad-based. Its principle states:

An organization should focus on what needs to be done—it should ensure commitment to EMS and define its policy.

This reflects one of the most important words in the ISO 14000 series: "commitment." Commitment to effective environmental management should be anchored to the organization's overall commitment to quality and innovation. Thus, it is clear that the first step in developing or improving an EMS system is to obtain commitment from the highest management level of

---

[46] The idea of "environmental profitability" has been used to describe the benefit of an EMS: Even though many significant effects of an EMS strategy aren't realized immediately, ultimately, an EMS will provide environmental profitability which can translate into overall profitability for the organization, profitability in terms of bottom-line numbers.

the organization, a commitment to improve the environmental performance of the organization in all phases of its business. This commitment as crucial.

The section is punctuated with "help" summaries and delineates the steps an organization might take to implement an EMS:

- Obtaining commitment of the highest management level to improve environmental performance;
- Completing an Initial Review of Environmental Position; and
- Developing an Environmental Policy.

Developing or improving an EMS may involve an initial environmental review. Such review identifies where an organization stands in terms of environmental protection. This review may cover a number of areas such as legal requirements; environmental aspects and issues; current management practices; past non-compliance; and the interests of internal and external parties.

Questionnaires, interviews, direct inspection, and a method noted as bench-marking (which represents a systematic comparison of key parameters and practices within the organization or in another organization) are among alternatives for conducting a review. Outside sources may include regulatory agencies, libraries; databases;[47] business associations; and professional assistance.

The review should note various operating conditions including possible emergency situations. The review should be documented.

Environmental Policy

An organization's environmental policy establishes the parameters for action and delineates the environmental performance required. It forms the basis for all future EMS activities. (To assist in focusing on general environmental concerns, Appendix A1.1. and Appendix A.1.2. provide examples of international environmental guiding principles from which an environmental policy can be formulated. The two examples cited are the Rio Declaration on Environment and Development and the International Chamber of Commerce's [ICC] Business Charter for Sustainable Development.) In

[47] Electronic information on the ISO 14000 standards is available through the Canadian Standards Association's (CSA's) World Wide Web page at : http://www.stoller.

broad terms, a business shall ascertain if the policy encourages: pollution prevention, reduction of both waste and consumption of resources; continued improvement; and the use of sound technology. The policy should reflect a commitment to these endeavors by top management and it should emphasize that the organization will comply with all applicable law.

The scope of environmental policy should take into account the fact that all activities, products, and services of an organization can have an environmental aspect which, in turn, have an impact on the environment. Policies must be developed to reflect an organization's commitment, minimize the impact, and protect the environment. Commitment will vary ranging from rethinking of product development to include all phases of the product's life from a natural state to it's disposal to prompting environmental education, and to encourage those with whom an organization does business to focus on sustainable development.

Planning

Before an organization can establish its policies and prioritize its objectives, environmental aspects and significant environmental impacts should be identified in order that potential risks are considered when setting objectives. Thus, the second principle identified is:

> *An organization should formulate a plan to fulfill its environmental policy.*

An organization should focus on what needs to be done and have a purpose and plan. The plan should consider as its basis the environmental policy. It may then focus on environmental aspects, objectives, targets, and legal requirements.

Although the Environmental Plan should include identification of environmental aspects, just what constitutes environmental aspects is open to interpretation. The Guidelines, Section 3.3, and ISO 14001, Section 3.3, define environmental aspects as "elements of an organization's activities, products, and services which can interact with the environment." It may be dredging; emissions from marine vessels; or storage of toxic wastes. Or it could include recycling of office paper or aluminum cans. If an environmental aspect causes a change in the environment, that change is referred to as an impact. If toxic wastes leak into the ground water or recycling of aluminum cans results in reduction of wastes, then the aspect has had an impact on the environment. An organization should implement and maintain a systematic approach to identify potential environmental aspects—benefits as well as negative impacts—in all phases of its business, its activities, as well as the materials it uses and produces. These should be identified and then prioritized. Areas that traditionally were not included in assessing environmental impacts such as marketing and profits may now identify these aspects. New and current operations and materials may be screened for environmental soundness. Procurement has an added environmental dimension. For many organizations, development of a database may prove valuable in establishing information to prioritize and assess. Also, the database would assist in environmental reviews.

Examples of questions to be considered in identification of environmental aspects and evaluation of environmental impacts, questions that, by their broad nature, are designed to trigger a full range of responses from all types and sizes of businesses include:

- Have potential environmental aspects of the organization's activities, products, and services been identified?
- Is the change to or in the environment as a result of the organization's activities (positive or negative), i.e., do they emit, discharge, or release anything into air, water or on land? Produce waste? Have contaminated sites? Store or transport hazardous or potentially problematic materials? The cultural as well as the ecological can be considered.
- Does the location of the organization require environmental considerations, i.e., is it located in an especially sensitive area such as near wetlands?
- How significant are the potential environmental impacts should a process failure occur?—"What if?"
- Does the organization train its employees to understand and deal with environmental aspects and impacts?
- Does the organization have a system to prevent negative environmental impacts from occurring?

Legal Requirements

Organizations should establish and maintain procedures to implement all environmental legal requirements which may apply to any of its activities, products, and services. Regulations may apply to specific activities such as site operating permits, authorizations, and licenses or transportation of hazardous substances; they may be industry-specific such as the international rules for oil tankers; and they may simply be general environmental laws such as the Endangered Species Act which may be triggered by a broad range of activities. Sources to identify and track such laws can include publications of all levels of agencies and industry associations as well as commercial computer databases such as Lexis and Westlaw. Professional legal services are another alternative.

Internal Performance Criteria

Internal priorities should be developed and implemented whenever there are no external standards or where they are inadequate. External and internal requirements are indicative of the performance necessary to meet an organization's objectives and targets. These can include a wide range of criteria such as management of air and water quality; contracts; regulatory requirements; and hazardous material management.

Environmental Objectives and Targets

Environmental objectives consist of the overall aim of environmental performance arising from the environmental policy; targets are the specific and measurable performance requirements that an organization sets out to achieve. Objectives and targets should be consistent with a business's environmental policy. Appropriate levels of management should set objectives and targets in light of relevant findings from environmental reviews, aspects, and impacts. Reviews should result in identification of specific, measurable environmental performance indicators which, in turn, can be used to either describe the organization's performance or to become the basis for objectives and targets.

Objectives and targets can be varied, from a commitment to reduce both waste and the consumption of resources to a commitment to promote environmental awareness. Objectives and targets can be written: 1) in terms of what is to be measured); 2) as a goal describing what an organization wants to achieve; and 3) as a time scale. Organizations should consider developing environmental performance indicators that can be self-assessed and may be externally verified and are comparable over time. Examples of indicators include: quantity of oil and its equivalent such as natural gas and coal used; amount of waste generated and recycled; water reuse; and dollars spent on protection of the environment.

Environmental Management Program

Organizations should implement programs to address schedules, resources, and responsibility for complying with their environmental policy and meeting objectives and targets. These should be revised to reflect changes in the organization. They may be free-standing or integrated into the organization's strategic plan. To begin the implementation of the Environmental Management Program, an organization should address such issues as prioritization of environmental issues; a cost/benefit analysis of options; identification of responsibility for implementation and provisions for post-implementation review; and assessment of adequacy of the program.

Implementation

This section is guided by the principle that:

> *For effective implementation, an organization should develop the capabilities and support mechanisms necessary to achieve its environmental policy, objectives, and targets.*

As an organization changes, so must an effective environmental management system adjust to these changes. It should also respond to the mandate of continual improvement.

Capability

Identification and availability of appropriate resources—human, financial, and technical[48]—are essential to an effective EMS and to meeting its objectives. Resources should be allocated based on a cost-benefit analysis.

Resources may be obtained in cooperation with outside organizations such as universities; Chambers of commerce; and other business organizations. Some state universities, for example, offer assistance in coordinating ISO 14000 training and workshops in preparation of third-party certification.

[48] For example, facilities and equipment.

EMS Alignment and Integration

Designing or revising the EMS so that it is integrated into the existing management system elements is critical. Management system elements that may require integration include the areas of policy, documentation, training, accountability, review and modification, and reporting.

Accountability and Responsibility

In light of legal ramifications stemming from liability for an environmental incident, considerable thought should be given to accountability and responsibility. In tandem with the assignment of responsibility for the EMS, it should come with sufficient authority and resources. Responsibilities should be clearly defined with all employees accountable in accordance with their responsibilities.

Included in considering accountability are the issues of performance review, training, reporting, and repercussion from non-conformance. The guidelines suggest an approach for developing environmental responsibilities which may be adapted according to the different organizational structures and the need to define environmental responsibilities based upon their own work processes. These are provided in terms of a sample of who may be assigned responsibility for what. For example, all staff must comply with defined procedures; and top level management establishes the organization's overall direction.

Environmental Awareness and Motivation

A common set of environmental values should be developed and reinforced taking into account views of interested parties. In focusing on these environmental values, it is incumbent that senior management play a pivotal role in communicating values and motivating employees. It should be ascertained how commitment to the enumerated policy is revealed; whether employees recognize this commitment; and how employees are recognized for adhering to the principles of environmental excellence.

Knowledge, Skills and Training

Enumerated objectives can only be met when personnel have appropriate knowledge and skills to effectuate the objectives. Thus, knowledge, skills, and training required to meet the objectives should be identified throughout the work-cycle of an employee beginning with recruitment. Employees should possess the requisite skills to meet their work requirements and possess an understanding of how their work may have an impact on the environment. The maxim holds true for contractors. They should ensure to the satisfaction of the organization that they are working within acceptable environmental parameters. Being current on regulatory requirements as well as internal standards is another priority. Training programs should address, among other areas, needs assessment; evidence of performance; and evaluation.

Support Action: Communication and Reporting

Reporting to interested parties on environmental activities of an organization provides evidence of an organization's commitment to environmental protection. Communication extends to external as well as internal parties and may describe an organization's environmental efforts to improve environmental accountability.

In order to continually improve an EMS, results from various monitoring activities should be given to appropriate personnel. Information sharing with other "interested parties" is also encouraged. Without information, it is difficult for personnel to act in an informed manner. The downside, of course, is deciding what information should be disseminated and to whom. (See Chapter Ten, "Legal Issues.")

Communication and reporting may focus on such areas as receiving and disseminating information; responding to internal and external parties; and communication to support continual improvement. Reports may include information on an organization's environmental policy, targets, and objectives. It is important that open communication be encouraged and that consistency in information disseminated be maintained.

The guidelines suggests ways in which an organization can communicate environmental information. These range from external routes such as annual reports, industry publications, and information hotlines to internal routes such as bulletin board postages and e-mail.

EMS Documentation

An operational process is another key to the success of an EMS. The emphasis is tri-fold: a) clearly defined procedures/processes; b) appropriately documented; and c) updated when necessary. The rationale is that documentation provides evidence of both what is required to achieve environmental objectives and evaluation of environmental performance.

The guidelines suggest organizations consider maintaining a summary of the documentation to serve as a reference. This summary may, among others, describe the means of achieving environmental objectives and targets and document the key roles, responsibilities and procedures. Chapter Seven, "Documentation, Document Control and Records," provides a comprehensive look at documentation, document control, and records.

Operational Control

Operational controls and procedures are needed to ensure that the level of environmental performance is consistent with the organization's policies, objectives, and targets. To establish and maintain these controls, the organization should consider various activities' significant impacts when developing or modifying operational controls and procedures. These may be identified through the various operational stages of an organization such as design, engineering, contracting, purchasing, storage, transportation, and marketing.

Emergency Preparedness and Response

Too often, time is not spent on the "what ifs" in environmental management. Emergency plans and procedures should be established to ensure adequate response to unexpected incidents. The organization should begin by defining and maintaining procedures for dealing with environmental emergencies.

The guidelines suggest what an emergency plan can include, ranging from a list of key personnel and details of emergency services to training plans. Some industries such as shipping are required to have not only an emergency response plan in case of oil discharge, but this plan must be approved by the U.S. Coast Guard.

Measurement and Evaluation

The principle for this section is:

An organization should measure, monitor and evaluate its environmental performance.

Measuring and Monitoring (Ongoing Performance)

Continual improvement can only be achieved through an on-going process of measurement, monitoring, and evaluation of actual performance in light of an organization's environmental objectives and targets. The guideline allows for "appropriate processes," not specific processes, which should be in place to ensure the reliability of data. The process includes evaluation of an organization's compliance with the law. An organization should identify what monitoring and control processes are in place and their effectiveness in terms of reliability.

Corrective and Preventative Action

Once the review and audit of the EMS is complete, the results—the findings, conclusions, and recommendations—should be documented. Equally important is the identification and implementation of the necessary corrective and preventive actions. Managers are charged with ensuring that these activities occur and that there is systematic follow-up.

EMS Records and Information Management

Records provide information on the relevant aspects of the EMS. The effective management of these records and other information is vital. Records should cover all the areas required by the standard. For a more thorough discussion of records and other documents, see Chapter Seven, "Documentation, Document Control and Records."

Audits of the Environmental Management System

An area which causes frequent consternation is that of auditing. Although the ISO 14000 series includes auditing documents, the Guidelines provide a brief background of "why," "when," "who" and "how": The *why*: To determine whether the EMS conforms to specifications and has been properly implemented and maintained. The *when*: They should be conducted on a "periodic basis." The *who*: by internal and/or external personnel who are objective, impartial and properly trained. The *how*: The EMS audit report should be submitted in accordance with the audit plan. Frequency of audits is determined by two factors: 1) the nature of the operation in terms of its environmental aspects and potential impacts; and 2) the results of previous audits. For further discussions, see Chapters Five, "Auditing Standards," Nine, "Self-Declaration and Third-Party Audits," and Ten, "Legal Issues."

Review and Improvement

The final principle holds that:

An organization should review and continually improve its environmental management system, with the objective of improving its overall environmental performance.

General

As with ISO 14001, the concept of continual improvement is a mainstay of ISO 14004. Continual improvement begins with measuring and monitoring.

Review of the Environmental Management System

At appropriate times, an organization should review the EMS to assess its effectiveness to ensure it addresses the environmental concerns of all its activities. The guidelines suggest a broad range of areas for review, from the findings of the EMS audit to an evaluation of the suitability of the environmental policy and the need for changes.

Continual Improvement

Continual improvement is achieved by constant evaluation of performance in light of environmental policy, objectives, and targets. The guidelines suggest that the continual improvement process should, among others, identify areas of opportunity for improvement and determine the cause of non-conformance.

Appendices

There are two appendices which provide examples of international environmental guiding principles: the Rio Declaration on Environment and Development; and the International Chamber of Commerce's Business Charter for Sustainable Development. The more salient Rio principles include:

- That "[i]n order to achieve sustainable development, environmental protection shall constitute an integral part of the development process and cannot be considered in isolation from it";[49]

---

49 The Rio Declaration on Environment and Development, Principle 4.

- That "States shall cooperate in a spirit of global partnership to conserve, protect, and restore the health and integrity of the Earth's ecosystem. In view of the different contributions to global environmental degradation, States have common but differentiated responsibilities. The developed countries acknowledge the responsibility that they bear in the international pursuit of sustainable development in view of the pressures their societies place on the global environment and of the technologies and financial resources they command...";[50]
- That "[t]o achieve sustainable development and a higher quality of life for all people, States should reduce and eliminate unsustainable patterns of production and consumption and promote appropriate demographic policies";[51]
- That "[e]nvironmental issues are best handled with the participation of all concerned citizens, at the relevant level. At the national level, each individual shall have appropriate access to information concerning the environment that is held by public authorities, including information on hazardous materials and activities in their communities, and the opportunity to participate in decision-making processes. States shall facilitate and encourage public awareness and participation by making information widely available...";[52]
- That "States shall enact effective environmental legislation, environmental standards, and management objectives and priorities should reflect the environmental and developmental context to which they apply. Standards applied by some countries can be inappropriate and of unwarranted economic and social cost to other countries, in particular, developing countries";[53] and
- That "States should cooperate to promote a supportive and open international economic system that would lead to economic growth and sustainable development in all countries, to better address the problems of environmental degradation....

50 *Id.*, principle 7.
51 *Id.*, principle 8.
52 *Id.*, principle 10.
53 *Id.*, principle 11.

- Environmental measures addressing transboundary or global environmental problems should, as far as possible, be based on an international consensus."[54]

The ICC's Business Charter for Sustainable Development begins by stressing that environment management should be a corporate priority: "To recognize environmental management as among the highest corporate priorities and as a key determinant to sustainable development; and to establish policies, programs, and practices for conducting operations in an environmentally sound manner."[55] Fifteen other principles address the areas of:

- Integrated management;
- Process of improvement;
- Employee education;
- Prior assessment;
- Products and services;
- Customer advice;
- Facilities and operations;
- Research;
- Precautionary approach;
- Contractors and suppliers;
- Emergency preparedness;
- Transfer of technology;
- Contributing to the common effect;
- Openness to concerns; and
- Compliance and reporting.

Each of these declarations provide general guidelines which may be of use to an organization that has little experience in formulating an environmental policy.

---

[54] *Id,* principle 12.

[55] ICC's Business Charter for Sustainable Development, Section 1.

5

ISO 14010, 14011, and 14012: Auditing Standards

I don't see no point about that frog that's any better n' any other frog.
Mark Twain, "The Celebrated Jumping Frog" (1867)

Introduction

The ISO 14000 series audit documents are designed as a basis for targeting areas for improvement: to assess: "How is the organization doing?"[56] The standards echo the common concerns of audits; that is, that they should be independent, objective, and accurate, and the results of the audit should be given to the organization who commissioned the audit. The documents are derived, in part, from the ISO 14011 series of quality system audit standards.

The ISO 14000 series has three auditing documents, all of which are guidelines but one of which provides general principles of auditing:[57]

ISO 14010: Guidelines for Environmental Auditing—General Principles
ISO 14011: Guidelines for Environmental Auditing—Audit Procedures: Auditing of Environmental Management Systems
ISO 14012: Guidelines for Environmental Auditing—Qualification Criteria for Environmental Auditors

---

56 An EMS standard provides guidance on establishing core elements of a system; an Environmental Performance Evaluation (EPE) standard provides guidance on how to measure performance on a regular basis against a set of objectives and targets established by EMS; and an audit standard provides guidance on how to verify the existence of an effective EMS against a set of agreed upon criteria.

57 More standards in this series may be prepared in the future.

Specifically, ISO 14010 provides general principles used in environmental audits and evaluates compliance of activities to environmental criteria. ISO 14011 assists organizations in developing and meeting the goals of their environmental policies as well as establishes audit procedures that provide for the planning and performance of an audit of an EMS to determine conformance with EMS audit criteria; it is applicable to all types and sizes of organizations operating an EMS. ISO 14012 provides guidance on qualification criteria for environmental auditors and lead auditors for both internal and external audits. Its informative Annexes describe the evaluation of the qualification of environmental auditors (Annex A); and Environmental Auditor Registration bodies (Annex B).

Audits do not go beyond compliance; at this point in the development of the ISO 14000 series, adequacy of such activities is not auditable. Functions performed by auditors under ISO environmental audit guidelines can be expanded at the express request of a client.

ISO 14010 defines environmental audit as the:

> *Systematic, documented verification process of objectively obtaining and evaluating evidence to determine whether specified environmental activities, events, conditions, management systems, or information about these matters conform with audit criteria, and communicating the results of this process to the client.*[58]

The following sections present highlights from each of the environmental audit documents. For a more thorough discussion of audits in general, refer to Chapter Nine, "Self-Declaration and Third-Party Audits."

ISO 14010: Guidelines for Environmental Auditing—General Principles of Environmental Auditing

The general principles of ISO 14010 are summarized:

- Types of environmental audits are not specified;

[58] ISO 1410, Section 3.9.

- Basic auditor qualifications include objectivity, independence, and competence;
- The auditing process involves due care, systematic procedures, confidentiality, and presenting the findings in a written report;
- The auditor and client (organization commissioning the audit) must agree on audit scope and criteria; and
- The audit report, in addition to the findings, may include such items as audit period, objectives, and identification of audit team members and audit representatives.

The series is intended as a guide for the general principles of environmental audits and the scope is broad. Audit-specific terms as well as general terms are defined. Audit-specific definitions include conclusion, criteria, findings, evidence and team, as well as audit, auditor, auditee, client, evidence, lead auditor, subject matter, technical expert, and verification. The client is the one who commissions the audit while the auditee is organization to be audited.

Only clearly defined and documented subject matter and responsible parties should be audited buttressed by appropriate resources. Without these, no audit should be attempted.

The General Principles (5.) include seven topics:

Objectives and Scope (5.1.)
Objectivity, Independence and Competence (5.2.)
Due Professional Care (5.3.)
Systematic Procedures (5.4.)
Audit Criteria, Evidence and Findings (5.5.)
Reliability of Findings and Conclusions (5.6.)
Reporting (5.7.)

The client should define the objectives of the audit while the scope is determined by both the auditor and the client. The team must be objective and independent of the activities they audit. This is important since the standards allow internal audits and self-declaration. To prevent a potential conflict of interest, if an internal auditor is used, she or he should not be accountable to those directly responsible for the area or function being audited. No matter whether the auditor is internal or external, the auditor should have the qualifications specified in ISO 14012.

The auditor should use reasonable diligence, care, judgment, and discretion. Confidentiality is mandatory: Environmental issues may give rise to potential liability. It is axiomatic, therefore, that the auditor or audit team never divulge audit information to any third party without proper approval of the client commissioning the audit and, where appropriate, the organization audited.

The audit should be performed according to the procedures delineated in ISO 14010 as well as ISO 14011. Although methods may vary according to the type of audit, unless there is some unique circumstance, consistency of procedures should be maintained.

Criteria should be determined as early as possible and communicated to the organization being audited.

The audit is meant to give a client being audited and the auditor confidence in the reliability of the findings. The organization commissioning the audit and, where appropriate, the auditee, should receive the results in a written report. Information to be included in the report is determined by the auditor in consultation with the organization commissioning the audit. Corrective action is not included in the report unless the inclusion of such information was specified by prior agreement. For a thorough review of the auditing process in general, see Chapter Nine, "Self-Declaration and Third-Party Audits."

ISO 14011: Guidelines For Environmental Auditing—Audit Procedures—Auditing Of Environmental Management Systems

The general principles of ISO 14011 are summarized as follows:

- Sample of typical audit objectives;
- Designation of responsibilities and activities of: lead auditor, auditor, audit team, client, and audit; and
- Auditing processes: scope, resources, pre-site visit activities (document review), audit plan, on-site activities, reporting, documentation, and record retention.

This document establishes procedures for conducting an environmental management system (EMS) audit, a process which should result in an organization meeting its environmental objectives. It is applicable to all types and sizes of organizations operating an EMS.

The Environmental Management System Audit Objectives, Roles, and Responsibilities section (4.) is divided into two main subsections: Audits Objectives (4.1.); and Roles, Responsibilities and Activities (4.2.). An EMS audit should have defined objectives such as determining whether the audited organization conforms to the EMS criteria. It should identify non-conformities to allow for improvement of the EMS. The document describes briefly the role of each participant in the audit process: the lead auditor; the auditor; the audit team; the client; and the auditee.

The lead auditor ensures that the audit is conducted within the scope of the audit plan, among other duties, auditors follow the lead auditor's direction, analyze and safeguard evidence, and assist in writing the report. The Audit Team should be selected on after considering a number of factors including ISO 14012 criteria, what is being audited, and special expertise of auditors such as technical and language skills.

The client's responsibilities and activities range from determining the need for and scope of the audit to providing appropriate authority and resources to conduct the audit. The organization being audited should, among other duties, provide the facilities and requested evidence, communicate to its employees regarding the objectives of the audit, and, of course, cooperate with the auditors.

The auditing section (5.0) is composed of four subsections:

Initiating the Audit (5.1.)
Preparing the Audit (5.2.)
Executing the Audit (5.3.)
Audit Reports and Records (5.4.)

The scope, determined by the client, the lead auditor, and generally, the auditee, provides the boundaries of the audit such as where the audit activities are to take place and on what activities the audit will focus. Resources should be adequate to ensure an effective audit.

The preliminary document review includes review by the auditor of such items as the organization's environmental policy statements, and other appropriate background information on the auditee's organization. The audit plan should be flexible to accommodate changes as evidence is produced. Team members should be assigned specific tasks such as focusing on a specific activity.

Working documents may include such items as checklists used for evaluating compliance with EMS objectives and records of reviews. Again, confidentiality, where appropriate, is stressed.

An opening meeting is designed to enhance communication and set the tone of the audit. It enhances the efficiency of the audit process through such procedures as introducing the participants, reviewing the scope of the audit, and highlighting the procedures to be used.

Methods of collecting evidence include interviews, verified by supporting information; examination of documents; and observation. Non-conformities should be documented.

All significant audit findings, and all significant non-conformities should be documented clearly with supporting evidence. These should be reviewed with the manager responsible for that area.

The closing meeting with the auditee prior to preparing an audit report; it is used to present audit findings to ensure the they are clearly understood. Final decisions of the findings rest with the lead auditor.

The lead auditor oversees the preparation of the audit report which includes both the findings and reference to supplementary documents. It should be dated and signed by the lead auditor. Other information may be included.

The lead auditor should send the audit report in a timely manner to those parties identified in the audit plan. Records should be retained as per the agreement among the client, the lead auditor, and the auditee.

ISO 14012: Guidelines for Environmental Auditing–Qualification Criteria for Environmental Auditors

The final auditing document provides qualification criteria for auditors. The general principles of ISO 14012 are found in the following sections:

Scope (2.)
Definitions (3.)
Education and Work Experience (4.)
Auditor Training (5.)
Objective Evidence of Education, Experience, and Training (6.)

Personal Attributes and Skills (7.)
Lead Auditor (8.)
Maintenance of Competence (9.)
Due Professional Care (10.)
Language (11.)

The document applies to both internal and external auditors.[59] Specific levels of education and length of work experience are required. For example, in general, auditors should have completed at least secondary education or equivalent and a minimum of five years appropriate work experience; auditors who have obtained a degree should have a minimum of four years appropriate work experience.

Auditors should have demonstratively completed both formal and on-the-job environmental auditing training. Formal training may include a focus on, among other areas, environmental science and environmental management systems. On-the-job training consists of a total of twenty work-days of auditing for a minimum of four audits within three consecutive years.

Among other attributes, an auditor should be able to communicate effectively, both orally and in writing; she or he should be objective, use sound judgment, and be sensitive to the business of social culture in which she or he is working.

In addition to the general attribute of auditors, a lead auditor must also possess leadership skills that would result in an effective audit. This may be endorsed by either interviews, observation and similar processes or participation in a specific number of hours of audits including in the capacity of acting lead auditor under the supervision of a lead auditor for a certain number of audits.

The auditor's knowledge should be up-to-date on matters such as environmental management systems; auditing procedures; and environmental law, science and technology. As well, auditors should exercise due professional care. Because effective communication is vital, audit team members should participate in an audit only when they are able to communicate effectively in the required language or are supported by a person who possesses the required language skills.

---

59 Although ISO 14011 provides some degree of information, requirements for collective qualifications of audit teams are not included.

Annexes

The ISO 14012 contains two informative annexes. The first, Annex A, relates to evaluating qualifications of environmental auditors and includes the evaluation process (A.2.) and the evaluation of education, work experience, training and personal attributes (A.3.); it provides a list of evaluation process methods.

Annex B relates to the environmental auditor registration body and contains briefly stated guidance on the development of a body to ensure consistency in the registration of environmental auditors. (*See* Chapter Eight for additional information on Accreditation Bodies and Registrars.)

An Additional Area of Focus

Several areas must be watched closely in order to ensure an effective environmental audit process. First, auditing standards must be consistent with the EMS standards. This would include consistency of third-party certification bodies and EMS audit findings. Secondly, because environmental audits are rather new in comparison with ISO 9000 audits and the basis for assessment is different, care must be taken to keep auditor judgments objective. And thirdly, with much of the ISO 14000 series still in the formative stage, steps must be taken to ensure consistency with national environmental auditing standards.

6
Implementation of ISO 14001

Never mind the quality, feel the width.
Vice Powell and Harry Driver,
Title of ITV comedy series, 1967-9

Reading the Standard

Level of Specificity

The ISO 14001 standard series provides both guidance and specifications. There are distinct levels:

- Those which are most restrictive - the "shall"-
  They are prominent throughout Section 4:

The organization *shall* establish and maintain a procedure to identify and have access to legal, and other requirements....

- Those which are less restrictive -"where/if/as appropriate" and "where practicable"- *If appropriate,* programs shall be amended....

"The organization shall also periodically test such procedures *where practicable.*"

"Records shall be maintained, *as appropriate* to the system and to the organization...."

- Those which are still important - "as required" and "in light of"

"... *in the light of* environmental management system audit results, changing circumstances...."

"...on the basis of education, appropriate training and/or experience, *as required.*"

For those "shall" requirements, the organization is free to choose how it must be done. This may be reassuring on the surface; however, the paucity of precise implementation methodology can lead to confusion. For example, Clause 4.1, Environmental Policy, states that:

> *Top management shall define the organization's environmental policy and ensure that it ... is documented, implemented, and maintained and communicated to all employees....*

The importance of this statement is not reflected by the fact that an organization's policy flashes on the monitor of every employee each time that employee switches on her or his computer but rather by the that the policy is *implemented, and maintained, as well as communicated.* Ensuring that this is done is the difficult issue and may be addressed through a number of alternatives such as workshops and training. A signed recognition by each individual employee that the policy has been read may not be enough.

Tackling A Clause

After reading a requirement clause, the question arises: "How may that requirement be implemented?" For example, Document Control, 4.3.5., requires that:

> *The organization shall establish and maintain procedures for controlling all documents required by this standard to ensure that...they are periodically reviewed, revised as necessary, and approved for adequacy by authorized personnel; the current versions of relevant documents are available at all locations where operations essential to the effective functioning of the system are performed.....*

Paraphrasing the clause does little to ensure that procedures are followed. An effective technique involves simply turning the clause into a question:

> *Procedures:* Does the organization have procedures for controlling all documents required by the standard?

If yes, the problem is solved; in most cases, organizations already have a record-keeping system in place but are the procedures up-to-date? If not, or if there are no procedures, then the organization must develop and maintain them.

Approval: How are the procedures reviewed and revised?
Who approves them?

The answer to these questions will allow an organization to implement the proper response to ensure that procedures are followed. This is not a call for complex layers, however. Concise and simple procedures will be, in the long run, the most appropriate path for compliance.

In tandem with this is the idea of inter-relatedness. Satisfying the clauses of ISO 14001 involves consideration of the interrelation among topic headings. It is this inter-relatedness that creates a unity of purpose for an organization rather than a modular approach in which independent units of an organization pursue unit-specific goals.

Process of Implementation

Although a management system may already be in place that addresses, in general, many ISO 14001 issues, several may require additional work. The following are ten steps to use as guidelines in the process of establishing an EMS.

Step One. Where to Start: Commitment

Commitment to effective environmental management must be anchored to the organization's overall commitment to quality and must emanate from the CEO or equivalent top-level management. Commitment translates into formulating a strong environmental policy statement; backing an environmental management strategy that focuses on environmental awareness and performance improvement; and buttressing it with necessary resources to implement activities that motivate, train, and keep personnel abreast of environmental aspects of their work. Evidence of this commitment should be seen in all sections, all parts, all units, and all functions throughout the organization. Top-level management should lead and participate in the initial policy-setting discussion forum. Summarization of the goals should be distributed throughout the organization.

Step Two. Review Where You Are

In beginning the process of ISO 14001 compliance, an organization would be well-advised to review the existing system and build on it. An organization may already have a management system in place through implementation of an ISO 9000 standard, through regulatory requirements, or through industry standards. For example, if an organization already has an ISO 9000 certified system in place, then many of the issues within an environmental management system such as document control and training issues have been addressed. For companies certified to ISO 9002, implementation of ISO 14001 may require pulling together what is already in place, expanding the policy, integrating the systems, identifying objectives and targets, and reviewing the document control procedure to see if exceptions need to be made. This initial review should result in an outline of: how an organization is currently environmentally positioned; how an EMS system may be managed; the degree of commitment by senior management; the resources available; and what operational standards have been set. This is not a life-cycle review; an organization need not evaluate each product, component, or raw material input. Rather, it is a selection of services, activities, or products to identify those aspects most likely to have a significant impact. The review may consist of "micro-EMS" implementation steps such as:

- Defining the objectives and parameters of the review;
- Identifying what resources are required: personnel (number, skills/technological, scientific and/or management background); technology, financial; and time allotment;
- Developing procedures for identifying, reviewing, and prioritizing environmental aspects;
- Deciding on whether any steps in the process—commitment, review, development, implementation, audit, continuous improvement— should include internal or external persons or both;
- Identifying regulations as well as any industry-specific environmental standards which should be noted or included in the development of the EMS; and

- Establishing an EMS plan indicating:
  - the organizational framework, *i.e.*, an integrated system;
  - the steps in the process; and
  - how each is to be addressed and by whom, *i.e.*, defining and assigning responsibilities.

An organization must implement and maintain a systematic approach to identify potential environmental aspects—benefits as well as negative impacts—in all phases of its business. Although an organization may already have focused on a number of environmental aspects, the review can identify areas which haven't been addressed such as marketing, finance, and procurement. Environmental aspects must be considered for all operating conditions: normal; shut-down, start-up, and emergencies. For many organizations, the development of a database may prove valuable in assessing and prioritizing information. This database could assist in future environmental reviews.

Prioritizing environmental aspects involves identifying potential risks through two or more levels of analysis. The simplest level could identify whether an organization's facility is on or near a water-body or a wetland; or on a geological fault line; or near a hospital or school. Another level would identify a potential hazard, i.e., the grounding of an oil tanker or the rupture of a pipeline. What the cost would be both financially (direct costs such as clean-up and remediation and indirect costs such as the sullying of an organization's image) and environmentally (natural resource impacts) should be considered.

Environmental laws and regulations encompass a great many environmental aspects. Therefore, an organization should consider any area, any item, on which law focuses. Aspects can be eliminated from the list only after determining the severity and duration of each aspect: How may it be controlled? It is temporary or on-going?

Finally, the question arises whether to use internal or external personnel to conduct the review. Use of in-house personnel who have experience in environmental matters brings to the table people who are already knowledgeable about the organization's structure and operation. External experts who have experience in developing an EMS come with proven ideas and options and may be able to view the organization from a more objective perspective. If outside personnel are used, however, they must be provided with the proper background materials for briefing and be placed in a situation wherein they are working in harmony with internal personnel.

Step Three. Development and Implementation of the Environmental Policy

Establishment of an environmental policy may be the most fundamentally important section of ISO 14001 in terms of the framework for implementing the standard. How objectives, targets, and implementation reflect the policy is one of the questions asked in an audit. The environmental policy must either be developed or approved by senior management. It is not set in concrete: as with the general philosophy of ISO 14001, it should be adaptable to meet the environmental and economic criteria of an organization, its technological options, and its financial, operational, and business requirements.

Although a verbatim quote is not necessary, the policy must include the underlying requisites presented in the standard such as a commitment to continual improvement, a commitment to the prevention of pollution, and a commitment to comply with relevant environmental regulations. If an organization already has an environmental policy, this may have to be modified to address specific requirements of the standard. This policy represents the environmental philosophy of an organization which should permeate all levels and functions of the organization. Ultimately, the efficacy of the policy will be tested through objectives and targets. Broad, unprovable tenets such as *"XXX will do everything it can to reduce emissions into the air"* should be avoided. *"Everything"* is unprovable. A more appropriate sentence may be: *"XXX will recognize and exercise its responsibility to prevent pollution at the source wherever and whenever possible; to conserve natural resources through the use of reclamation and other appropriate methods; to continually improve its performance through training, audits, and reviews; and to assure that it meets the requirements of all applicable law."*

The standard also requires that the policy must be made available to those outside the organization: neighbors, investors, shareholders or simply those making a research inquiry. In other words, the policy must be reasonably accessible.

Step Four. Know the Applicable Law

This step is simple on its face. The larger and more complicated an organization, the more resources must be dedicated to fulfilling this mandate. Requirements may be found in federal, state, and local laws, regulations, rules, industry standards, and company policies and procedures. Multi-national organizations require more extensive reviews when the law of other nations are complex.

Step Five. Identify Objectives and Targets

Measurable objectives and targets to reduce negative environmental aspects must be delineated. These must be consistent with the policy and take into account the views of interested parties. The list of who may be "interested" is far-ranging. It could include, among others, customers, suppliers, shareholders, investors, insurance companies, environmental organizations, the community, financial institutions, law-makers, contractors, and mass communicators. Thinking about objectives from an interested party's perspective gives an organization a broader and perhaps unique view of environmental issues.

Specific objectives such as *"recycling 50% of waste in year X"* is acceptable; general concepts such as *"reducing pollution"* will not satisfy the requirements of the standards. The objectives and targets must be documented and maintained for each relevant function and level.

Step Six. Spread the Word: Say What You Do

The theory behind an EMS can be stated simply: Say what you do, do what you say, and prove it. Successful implementation of an effective EMS depends on ensuring that all personnel understand the significance of an EMS and their individual role in making it successful. Motivation will be brought about only with a basic comprehension of environmental consequences. Persuading people to think about environmental impacts is one of the more challenging steps.

Senior management should look at the organization as a whole and at each of its parts to discern how the environment may be affected by what it does, what it produces; and then think of stewarding products and services as well as seeing product design as "circular." The standard maintains that senior management play a key role in motivating employees. The motivation of personnel has been the topic of countless books and journals but what is important in dealing with the ISO 14000 series is proving that all members of the organization understand the importance of achieving the environmental objectives and targets for which they are responsible and/or accountable. It should be demonstrated that personnel were encouraged to achieve these goals. Examples of meeting this challenge include environmental awareness workshops and education seminars that stress not only the soundness of having an effective EMS, but how an EMS affects the organization and its personnel; what advantages would be derived in both the short and long term; and what benefits, including indirect benefits, will ultimately strengthen an organization and that which surrounds it—the neighborhood, the area, the community, and the state.

Motivation to continually improve may be enhanced when employees are recognized and rewarded for achieving environmental objectives and targets. However, it is important to distinguish perceiving rewards for reward sake. The latter would obfuscate the underlying purpose of an EMS; the value of rewards should emanate from what is achieved, directly and indirectly, through an effective EMS—a cleaner and protected environment.

Employees should be encouraged to make suggestions that can lead to improved environmental performance. Creative problem solving by the people who are in the forefront of the process, the front-liners who face a problem squarely, everyday, can be an effective way to generate solutions for troublesome environmental issues.

Step Seven. Do What You Say

Focusing on environmental objectives and targets in every relevant part of an organization may be new to many organizations. A program to meet objectives and targets at each relevant function and level of the organization must be established and communicated; the program should specify action steps, schedules, resources, and responsibilities required to achieve objectives and targets.

Management is required to demonstrate commitment to an EMS. A management representative must be charged with the responsibility to oversee the EMS and report on its status. In line with this is allocation of required resources for maintaining an effective EMS.

Step Eight. Prove It

The ISO 14001 standard requires that proof of an EMS in conformance with the standard be documented. At this point, a decision must be made on what method of controls will be used to ensure compliance with the standard as well as a determination of what must be documented.

Proper environmental documentation ensures that people know who is responsible for what; this obviates the possibility that something is passed over simply because no one knows who is responsible for a task. Documentation and audits also provide a mechanism for analyzing what was done right and identifying what needs to be corrected.

Audits may be completed by impartial external or internal persons; self-declaration is a method of gaining ISO 14004 certification to the EMS standard as is third-party audits. An organization's internal auditors immediately offer in-depth knowledge of the unique inner workings of an organization while third-party audits offer a high degree of unbiased observation and may be particularly appropriate for those complex organizations and/or those involved in global markets in which effective environmental management systems are a prerequisite to doing business. Self-declaration may be more suitable for a smaller organization whose long-time associates would accept such declaration. Although certification details are not provided in ISO 14001, the third-party audit can include similar steps as those involved in an ISO 9000 standard audit through the guidelines set out ISO 10011-1 and 10011-2 as well as the general audit guidelines in the ISO 14000 series.

Step Nine. Satisfied? Management Review

An EMS is a living system with a goal of continuous improvement. This can be achieved only when the EMS is reviewed at appropriate intervals by senior managers for suitability, adequacy, and effectiveness. The review emphasizes the continued commitment by top-level people. Objectives and procedures should also be reviewed by the level of management that defined them.

Step Ten. Continual Improvement

The last step is not the final step. ISO 14001 standards' conformance can be assessed from the corporate level to the site level and it should be reviewed frequently to improve and expand the scope of assessment. Continual improvement is of high priority; mechanisms to encourage continual improvement such as "environmental thinking" workshops should be considered. Seminars and other lines of communication should be created so that ideas are exchanged and personnel at all levels share the commitment to environmental excellence and continual improvement. Personnel should be encouraged to find ways to eliminate sources of liability and unnecessary cost. Cost reduction can be realized even through simple steps: recycling paper for interoffice memos and tablets; reusing interoffice envelopes; recycling aluminum soft drink cans and tin cans; using white-only tissues and paper towels; purchasing only non-toxic board markers and re-capping the tops; copying on both sides of paper; eliminating loss of energy from cold stores when doors are opened and closed; promoting commuter cooperatives; reducing electrical power used in office lighting; and encouraging environmental cost reduction alternatives to be used by contractors.

Conclusion

Although implementation may be handled in a myriad of ways, some organizations find it beneficial to divide the standard's criteria between quality assurance and environmental coordinator personnel. A management system in general may already focus on such issues as consistency, corrective action, document control, and continuous improvement. Oftentimes, this pre-existing management system has been developed to the point where implementing many of the ISO 14001 provisions involves little more than "filling in" and improving record-keeping and documentation methods. In other words, an internal assessment of the progress of an organization for each requirement is followed by a "gap analysis" to dictate where improvement must be made. Building on an already existing system may only require modification of operating procedures, for example. What should be taken into consideration in developing an ISO 14001 environmental management system is that integration is tied to both operational and strategy considerations. Not only is it more cost-effective to have one management system that includes environmental management, but it also demonstrates that environmental stewardship is part of the over-all philosophy of an organization's management effort.

7
Documentation, Document Control, and Records

The volume of paper expands to fill the available briefcases.
Jerry Brown

Documentation

The ISO 14001 standard was developed as a mechanism to provide evidence that an organization is committed to an effective environmental management system (EMS) for environmental protection. This requires evidence that an EMS system is established, fully implemented, and documented where required. An organization will be in compliance with ISO 14001 when all applicable requirements of the standard are met. The standard must be read carefully to garner what documentation is required. In some cases, the standard specifies that documented procedures covering a clause, in whole or in part, are a requirement. In other clauses, mere evidence of established procedures satisfies the standard.

Documentation of the System

The ISO 14001 standard requires proof that an organization's environmental management system (EMS) is in effect through a written description of the EMS. Documentation must be made of such items as the EMS policy; objective and targets; core elements; directions for documentation; the audit; who is accountable and for what; operational control; monitoring and measurement of significant environmental impacts; and corrective action.

The standard includes clauses which require "establishment" but do not specifically ask for documentation.

Documentation of Implementation

The specification standard also requires evidence of implementation of the environmental management system. These records demonstrate that procedures have been carried out to ensure the system is effective. These relate to management review; identification and maintenance of environmental records; procedures for ensuring compliance with environmental law; communication from external parties about the EMS; and other measuring activities such as conformance with the EMS objectives.

In general, documentation of the system describes the "what's" while documented procedures state strategies and responsibilities, the "how, when, where, and who." One method of documenting procedures is to examine the functional areas of an organization, with each manager accepting responsibility for documents relating to those areas under her or his control. Responsibility includes reviewing, approving, and ensuring that procedures are communicated and understood by appropriate personnel.

All procedures are not driven by specific law; procedures may require establishing evidence of or documenting common sense. For example, an organization may not have documented or have evidence of an established procedure for how much soil to use in covering non-toxic refuse at a waste facility, and yet, this may be a procedure which requires documentation. Various options make documentation more effective:

- Using of flow chart processes before beginning to write procedures;
- Writing with document control requirements and audits in mind;
- Including paragraph cross-referencing;
- Being consistent throughout documentation;
- Using section headers;
  - document number
  - revision level/status
  - approval signature
  - page __ of __
- Using job titles rather than names of personnel;
- Defining special terms;

- Not including proprietary information.

"Procedure" content may include the:

- Purpose/intention of the procedure;
- Scope outlining the area, department, group, personnel, or circumstances to which the procedure applies;
- Responsibility, the person or functional area who has authority to perform and ensure performance of the procedure;
- Definitions;
- Action or procedure statements detailing the action of those personnel involved in the activity; if possible they should state who does what and how, as well as where, when, and why the activity is carried out; and
- References to other documents which could have a bearing on the activity documented within the procedure.

The format should be such that the contents are easily modified as an organization changes. It should be clear and to the point.

Document Control

The ISO 14001 specification document also includes major reference to documentation control. An organization is required to establish and maintain procedures for controlling all documents required by the standard to ensure that they are locatable; located where appropriate; approved for adequacy; revised when necessary with each revision dated; and current with obsolete documents removed unless identified and retained for a particular purpose. Requirements also include proper storage to preclude damage or loss as well as procedures required for such storage. As well, documents may be in a computer-based format.

> The Informative Annex to ISO 14001, Annex A, offers additional guidelines for documentation concerning the level of detail of documentation. It should allow a user to find additional information on various sections of the EMS. However, a single EMS manual is not required.

If a single EMS manual is used, it can enumerate requirements and responsibility for document control. There are a number of items; including several of these previously noted, for possible inclusion:

- Records should be maintained to demonstrate the organization's compliance with the EMS;
- All documents associated with the EMS should be formally controlled including maintaining procedures for identification, collection, indexing, filing, storage, maintenance, disposition and retention;
- Storage conditions should be adequate to preclude loss or deterioration and damage;
- The organization should seek to prevent unnecessary duplication of records and types of records;
- Control of all distribution is the responsibility of the EMS Coordinator;
- A list of all document control procedures should be maintained by the EMS Coordinator; document control procedures must be available at relevant locations;
- The EMS Coordinator is responsible for updating and replacing requirements as necessary; and
- There should be a description of collection, indexing, and filing procedures.

A problem may arise from not knowing just how much and what to write in each instance where documentation is required from both a practical and operational perspective as well as from a legal perspective. (For a discussion of Legal Issues in Environmental Audits, Chapter Ten.) From an operational perspective, simplicity is the key. One of the most effective ways of addressing each of the requirements is by re-phrasing them into a series of questions. To exemplify, the issue of training is used:

- **Question One.** What is the level of the requirement? A "shall"? or an "if appropriate"?

The organization *shall* identify training needs.

Thus, training is required.

- **Question Two.** How does the organization currently deal with the issue?

The organization must identify appropriate training, if any, currently in place.

- **Question Three.** Must records be kept of training?

The organization shall ... and maintain procedures for the ... maintenance, of environmental records including training records.

- **Question Four.** Who in the organization can address the issue of training records and their retention?

- **Question Five.** If there is no training or no records, who can implement this requirement?

Document control confirms that all personnel are working from the latest revision of any procedure. Without this control, an EMS is not effective. Organizations should maintain documents in a manner sufficient to implement the EMS. At the same time, however, the focus should be on effective environmental protection; this should not be overshadowed by a complex array of documentation requirements.

Records

Specific records provide evidence of an organization meeting ISO 14001 requirements and a functioning EMS. They also provide evidence of deficiencies or problem resolution as well as traceability and status.

An organization should describe:

- What records are to be kept.
- How and where these records should be kept.
- The retention time for the records.
- How the records should be controlled.
- Who is responsible for the disposition of the records.

An auditor would also address such issues as:

- Retrievability: records should be identified, collated and stored so that they are readily retrievable;
- Legibility: records must be legible; and
- Protection: records must be protected against damage, deterioration, or loss.

Environmental records can include compliance records; incident reports; training and audit records; management reviews; contractor/supplier information; and maintenance records.

Conclusion

ISO 14001 does not require an organization to have an EMS manual. But it does require that an organization establish and maintain information describing the core elements of the environmental management system and their interaction, and provide direction to related documentation. Because each organization is unique, there is no standard prototype EMS: The EMS must describe an organization's operation and accurately reflect the system which is operating. Any method in compliance with the ISO 14001 standard demonstrates a commitment to environmental protection.

8
Accreditation and Registration

Ah, what a dusty answer gets the soul when hot for certainties in this our life!
George Meredith

Criteria for Registrar/Certification Bodies[60]

To attain a level of "compliance certainty" is one of the objectives of third-party registration (or certification, as it is referred to in the United States.) A number of benefits can be derived from third-party registration of an organization's environmental management system (EMS). The primary benefit is validation by an outside agency that the implemented EMS meets the requirements of the ISO 14001 standard. This type of recognition provides all those internally involved in implementing and maintaining the system with an acknowledgment that an important goal has been achieved and the effort and resources expended are valued and recognized. Other benefits are derived from the potential marketing opportunities enhanced through use of the registrar's reputation and credentials. Regular visits by the registrar encourage continued compliance and foster an atmosphere in which the continuous improvement process can flourish.

Almost anyone can register an organization or issue certificates, but not all certificates are recognized by a national or official government agency. In

[60] According to ISO/IEC Guide 2: General Terms and Their Definitions Concerning Standardization and Certification, certification is a procedure by which a third-party gives written assurance that a product, process, or service conforms to specified requirements. Registration is a procedure by which a body indicates relevant characteristics of a product, process, or service, and then includes or registers the product, process, or service in a publicly available list. Accreditation is a procedure by which an authoritative body gives formal recognition that a body or person is competent to carry out specific tasks. Unfortunately, certification is often used interchangeably with registration in the U.S. while in Europe, the term used for certification of a system is environmental management system registration.

order for a certification body to issue ISO 14001 certificates of conformance, that body should itself be accredited by an official agency. Thus, the national accreditation body accredits the certification body. A simplistic view of how the process works is illustrative:

- A nation creates national accreditation bodies;[61]
- These bodies accredit registrar bodies;
- The registrar audits organizations to assess conformance with a standard; and
- The registrar may certify that the organization is in conformance with the standard.

For the ISO 9000 series accreditation, three major bodies serve the North American market: the American National Accreditation Program for Registrars of Quality Systems in the United States;[62] the Raad voor de Certificatie (RvC) in the Netherlands; and the National Accreditation Council for Certification Bodies (NACCB) in the United Kingdom. The RvC accrediting body accredited U.S. registrars such as ABS Quality Evaluation, Inc. In the United States, a subsidiary of the American Society for Quality Control (ASQC)—the Registrar Accreditation Board (RAB)—has implemented an accreditation program for quality system certifiers.

The American National Standards Institute (ANSI) has not decided whether it should offer ISO 14001 accreditation services independently or in conjunction with other organizations.[63] Accreditation services include accreditation of registrars and accreditation of course providers. ANSI, the Registrar Accreditation Board (RAB), and the Environmental Auditing Roundtable (EAR) have presented their qualifications to be the U.S. ISO 14001 accreditation body to the ANSI's Board Committee on Conformity Assessment (BCCA).

---

61 In Europe, there is one accreditor per country but multiple registrars per country.

62 The National Program for Quality Registrar Accreditation was formed in 1991 by the joint effort of the American Society for Quality Control's (ASQC) Registration Accreditation Board (RAB) and the American National Standards Institute (ANSI). RAB administers the overall operations of the program. ANSI is the coordinator of the U.S. private-sector voluntary standards system and the official U.S. member body to the ISO and the International Electrotechnical Commission (IEC), via its U.S. National Committee.

63 As of September, 1995.

Auditor Registration

Auditor registration was one of the most hotly debated issues within the negotiation of the ISO 14000 series auditing documents. It was felt by some that registration was not an issue delegated to TC 207, the Technical Committee who developed the standards, and should, instead, be left to the individual nations and their representative registration bodies. Others argued that the registration guidelines were generic enough to be acceptable. Ultimately, Annex B, "Environmental Auditor Registration Body," was included in the final draft. It contains only two sections:

- **Clause 1:** *B.1 General*

This clause describes the rationale of the annex: To provide consistency of registration of environmental auditors.

- **Clause 2:** *B.2 Auditor Registration*

The annex does not require that a body for accrediting environmental auditors be established; but when it is appropriate, the body must ensure that environmental auditors are registered in a consistent manner and that such a body should be independent and follow the annex's guidelines:

The Choice

Assuming that an organization chooses to use a third-party auditor to obtain certification of conformance to ISO 14001, a number of issues arise, not the least of which is what registrar is the most appropriate choice for a particular organization. An organization should consider where the registrar and its certificate are recognized. For example, if the organization primarily exports to the Netherlands, then the registrar should be recognized by the Netherlands's national accrediting body. Once a list of accredited registrars has been gathered by the organization, an organization should further evaluate potential registrars by considering credibility, competency, consistency, and cost.

Credibility. Does the registrar have a stable history with evidence of the resources and business activity necessary to sustain it as a viable business during what will be a long- term relationship with its customer? Is the registrar accredited by at least one well-recognized accrediting body? Is this accreditation granted to the office from which an organization's program will be managed or is it only available through the registrar's sibling or parent company?

Competency. Is the registrar accredited to operate its program in an organization's particular industry? This information may be obtained upon review of the registrar's accreditation certificate and the certificate appendix listing industries that fall within the registrar's accredited scope. If an organization's particular industry is not included in the registrar's accreditation, then the certificate issued will not include the credentials or marks of the accreditation bodies. Specific industries are included in a registrar's accredited scope when that registrar has demonstrated to the accrediting body that its audit personnel have the industry experience and qualifications to conduct effective audits in that industry.

For ISO 14001, an issue is whether the auditors must have specific competence within the activity or industry being audited or whether a general background is sufficient. The complexity of the environmental arena may demand specificity even though the ISO 14001 standard itself is generic in content. Under ISO 14001, an organization must maintain a procedure to identify the environmental aspects of its activities, products, and services. The auditor may immediately face the conundrum of how "establishing and maintaining environmental aspects" is satisfied. Two different organizations may conform by completely different performances. Whether having experience in an industry lessens the difficulty is problematic for there are those who feel that technical competence in a specific area is not necessary; that, in fact, a more objective audit can be completed when an auditor is not saddled with specific environmental preconceptions but instead possesses general environmental knowledge.

The qualities of an auditor are specified in general terms in ISO 14012. It describes educational and work experience requirements, auditor training, personal attributes and skills, maintenance of competence, due professional care, and language requirements. In Annex A of ISO 14012, guidelines for evaluating the qualifications of environmental auditors are suggested. In general, an environmental auditor must comprehend environmental issues. Understanding of formal management systems is not enough to assess compliance in the environmental arena. Unlike a management system with broad-based objectives, some of which may be vague, an ISO 14001 environmental management system (EMS) requires that objectives and targets be established at all relevant levels; thus, an auditor must search through these operational levels to ensure that the goals have been met.

Of particular challenge for auditors is the standard's requirement that all relevant laws be adhered to which includes the applicable law of every state and nation in which the organization does business. The result is that an auditor must look for evidence that an organization has identified all applicable law, regulations, and rules. The auditor will look for evidence of how an organization decides on the applicability of each relevant law; a superficial overview will not provide the necessary assurance. For organizations doing business in several countries, an auditing team may incorporate an auditor who is familiar with compliance issues of a nation's specific law.

Consistency. Too often, lack of consistency in audit conduct and audit findings is prevalent. No registrar can absolutely guarantee that its auditors will see the same things in the same way each time. After all, an audit is conducted on a sample randomly drawn from a defined population and results can differ if the samples differ. However, the registrar should have established procedures and training programs to enhance consistency in application of the standard's requirements and a defined appeals process as safeguards for customers. An organization should note how the registrar assures consistency in application of the standard from audit to audit and customer to customer; for example, if the registrar uses subcontractors, an organization should inquire as to what procedures are followed to ensure consistency.

Cost. The time it takes to implement an environmental management system depends on the organization's current status, its commitment to the implementation of an EMS, and the resources it is willing to expend to implement the system. The time for actual certification is dependent on the preparedness of the organization and the number of deficiencies detected during the audit.

There are many costs associated with certification, the first of which is the cost of actually developing and implementing the EMS. A organization may elect to use only internal resources to implement the quality system or it may elect to rely completely on the services of an outside consultant; or it may incorporate a combination of both approaches. There is no guarantee from any registrar that if a consultant is used, or a system is implemented internally, or a combination of the two is adopted, that certification will be achieved.

Assessing the costs of the actual certification process should be done in great detail. It is important when selecting a registrar that a organization be totally familiar with all costs associated with the certification program. The organization should assure that it has obtained a price for document review; a price for a pre-assessment visit if one is deemed necessary; the cost of the actual assessment; and any costs associated with issuing the certificate and writing the report. The organization should know the policy of the registrar relative to surveillance visits including the duration and cost of those visits. The organization should know the validity length of the certificate as well as any reassessment or partial reassessment costs which may be required in the future. Furthermore, an organization should be aware that some registrars require application fees, listing fees, and certification fees in addition to those prices normally quoted with the above actions. An organization must ask the registrar about all fees to assure the full cost of the certification process is known.

In summary, the registrar must be selected on the basis of an organization's needs. To find out about the ability of a registrar, an agent should contact more than one and ask questions about how they operate, their scope, their capabilities, and all costs. It should focus on credibility, competency, consistency, and cost.

9
Self-Declaration and Third Party Audits

There are only two qualities in the world: efficiency and inefficiency, and only two sorts of people: the efficient and the inefficient.
George Bernard Shaw, "John Bull's Other Island" (1907), Act 4

Should There be an Audit?[64]

The decision to conduct an audit is a crucial one. Incentives may include, among others, avoidance of liability and associated penalties; cost efficiency; improved risk assessment and management; reduction of accidents/incidents; requirements of customer; Security and Exchange Commission (SEC) regulations requiring disclosure of environmental liabilities; improved public perception; and requirements of financial and insurance institutions and contracts. A compliance audit may be used to mitigate criminal sentences and is one factor considered by the EPA and the U.S. Department of Justice (DOJ) in determining whether to prosecute. The DOJ considers: 1) whether the company has regular procedures, including audits, to evaluate, detect, prevent, and remedy incidences which may lead to non-compliance; 2) whether such audits are conducted; and 3) whether the auditor's recommendations were followed in a timely fashion.[65] Environmental compliance audits are also supported by the U.S. Sentencing Commission's "Draft Federal Sentencing

---

[64] An audit may be an "environmental assessment" which is typically associated with property transactions, or an "environmental compliance audit" which is used to assess compliance with law and policy applicable to businesses.

[65] U.S. Department of Justice, *Factors in Decisions on Criminal Prosecutions for Environmental Violations in the Context of Significant Voluntary Compliance or Disclosure Efforts by the Violator*, 1 July 1991.

Guidelines Related to Environmental Crimes."[66] One thing is certain: The complex technical and legal content of environmental law makes full compliance impossible and makes compliance assurance difficult for even well-staffed companies. Multi-million dollar fines or penalties or jail terms are associated with even technical violations. Audits identify areas for non-compliance and serve public and business purposes by helping companies pro-actively achieve regulatory compliance.

Because the results of audits may expose an organization and its employees to civil and criminal liability, before conducting an audit, an organization should be prepared to remedy any identified problem. Preparation should include adequate financial and personnel resources in order to act quickly if non-compliance is found. Recurring or costly non-compliance problems should be addressed through an action plan to identify the basic cause of the problem and reduce recurrence. Corrective action plans are indispensable; to voluntarily audit and not incorporate such plans simply begs for additional difficulties.

This evidence is required in order to continually improve an EMS. Audits reveal the good, the bad, and the ugly; the information in an audit report is essential to management in making decisions. Results from monitoring, audit, and management review should be communicated to those within the organization who are responsible for improvement. Information-sharing to other "interested parties" is also encouraged. The rationale is that it encourages public understanding of efforts to improve environmental performance. Without information, it is difficult for a party to act in an informed manner. The downside, of course, is deciding what information should be disseminated and to whom. The potential for liability is too strong to suggest a *carte blanche* approach, while trickles of information can severely damage the effectiveness of an environmental management system (EMS) and at the same time negate the spirit of ISO 14001.

---

[66] *See* United States Sentencing Commission, Amendments to the Guidelines for United States Courts, Section 8A1.2, Comment(k), 56 Fed. Reg. 22762 at 22766 (16 May 1991); 16 November 1993 Draft Corporate Sentencing Guidelines for Environmental Violations Developed by the U.S. Sentencing Commission, printed in Envt. Rep. Curr. Devl. (BNA, Washington, D.C.).

Types of EMS Certification Audits

Under ISO 14001, an organization must establish and maintain a program and procedure for periodic environmental management system audits to be carried out in order to ascertain conformance with the EMS plan and provide feedback to managers for review and continuous improvement.

The ISO 14000 series audit documents (ISO 14010, 14011, and 14012) address the concerns of auditors rather than clients. The emphasis is on procedures and documentation, including audit reports, which oftentimes translates into increased costs with "added value" to the audit process. The audit is not intended to evaluate the effectiveness of an EMS. That is a management decision and should not be subjected to the vicissitudes of an auditor.

Certification to ISO 14001 may be accomplished through either an audit resulting in self-declaration or a third-party audit. Self-declaration holds the organization out as having fully implemented ISO 14001 which may include internal audits of the systems using the ISO 14010, ISO 14011, and ISO 14012. A third-party audit provides independent verification that an organization's EMS is in conformance with ISO 14001 and may also incorporate the ISO 14010, ISO 14011, and ISO 14012 documents.

Self-Declaration

The need for environmental self-audits is widely recognized.[67] In organizations with an effective EMS, non-compliance findings will be remedied quickly as non-compliance is immediately brought to the attention of management. In the initial audit it may be appropriate to focus on the most critical section of an organization's operation, *i.e.*, waste disposal, and apply the standard to that section. Frequent audits would be developed based on that information. Other areas of importance should be identified and prioritized.

Self-declaration may be more appropriate for a smaller organization whose long-time associates would accept such declaration. Also, an organization's internal auditors immediately bring to the table an in-depth knowledge of the unique inner workings of an organization.

---

67 See, for example, Weinstock, Matthew, "Environmental Auditing: A Measure of Safety," *Occupational Hazards 73* (May 1993).

Third-Party Audit and Registration

Who benefits from third-party registration? Registration may result in a useful marketing tool: Third-party assessment offers a high degree of unbiased observation and may be particularly appropriate for those large, complex organizations and/or those involved in global markets in which effective environmental management systems are a prerequisite to doing business. It certainly may help regulators—the EPA's and the OSHA's of the world—to re-focus their attention away from registered entities and focus on those who may be less circumspect. This translates into fewer inspections and fewer pieces of paper required to be submitted for registered companies. But first, it must be determined how the EPA or equivalent regulator views the ISO 14000 series: If it grasps its tenets wholeheartedly, then either certification process—third-party or self-declaration—may be acceptable.[68] The EPA has a natural interest in environmental management systems in general and the ISO 14000 series in particular: The ISO 14000 system can be used for compliance assurance and may go beyond compliance; and the concept is utilized in EPA's Common Sense Initiative (see Chapter Two, "Guidelines for Environmental Management Principles and Supporting Techniques"). An organization may receive fast-tracking for permits, Preferred Government Supplier status, and recognition of due diligence resulting in potential lower penalties for non-compliance. At the same time, if all the competitors are certified, then there is no additional advantage except that an organization has kept up with its competitors.

Audits in General

No matter which is chosen—third-party or self-declaration, all audit experiences should reflect a team concept, one in which the players have a common goal: to effectively manage those aspects of an organization which may have an impact, directly or indirectly, on the environment. Thus, an auditor must approach an organization with a sense of professionalism that includes an enthusiasm for the mission and a responsibility to assist in ensuring that there is improvement.

[68] Self-declaration could be accommodated through the EPA finding alternative EMS training certification programs for small businesses that cannot afford third-party assessment.

An organization's personnel must understand environmental issues as they relate not only to the technical side of the organization but as they relate within the management scheme itself. Further, it is incumbent upon managers to give full support to auditors and for that spirit to be reflected throughout all aspects the organization.

In general, audit programs should cover:

- Scope—the actives and areas to be considered;
- Frequency;
- Responsibilities and requirements associated with conducting audits;
- Communication of audit finding;
- Auditor competence; and
- Methodology—how the audits will be conducted.

Audit Procedures[69] and the Registration/Certification Process

General procedures for third-party audits include:

Pre-audit

An auditor will review the EMS documentation. In most cases, the organization will submit a manual or set of manuals and the auditor will perform a desk audit in which the documents are compared to the requirements of the standard. The auditor will usually communicate with the client to ask for clarification if necessary or inform the client that the policy statements either meet or do not meet the intent of the standard. It is important to provide briefing documents to auditor(s) and allow adequate time for review. What documents an auditor might review in advance include:

[69] Proper auditing techniques may be found in such publications as Environmental Auditing Rountable Procedures Work Group, "Guidelines for Written Environmental Audit Program Procedures," 1990; Greeno, L.G. Hedstrom and M. DiBerto, Center for Environmental Assurance, Arthur D. Little, Inc., "Environmental Auditing, Fundamentals and Techniques" 1985; Smith, A.C., "Continuous Improvement Through Environmental Auditing," in *Total Quality Environmental Management,* Winter 1991/2, pp. 121-129; Smith, A.C. and W.A. Yodis, "Environmental Auditing Quality Management," Executive Enterprises Publications, 1989.

- Company annual reports (present and prior three years);
- Company sales literature;
- Regulatory violation reports;
- Training/Educational records;
- Safety Records;
- Summaries of regulations;
- Industry standards;
- Permits; and
- Better Business Bureau and media reports.

Audit

Basic agreement on the scope and audit procedures should be established initially. Certification of only certain sections, parts, functions, or aspects of an organization is possible but this approach contrasts with the intent of ISO 14001: to have an integrated environmental management system. Without integration, theoretically, a company could operate in the U.S. with one standard and have no standards in, say, a lesser developed country.

Housekeeping procedures must also be addressed. For example, may cameras be used by the auditor? Photographs provide an excellent mechanism for recording non-compliance. Conditions for its use may include that copies of all negatives and/or photographs must be left with the organization after the audit. What if it rains? Procedures should be established in the event of inclement weather. Has a room been reserved for the lead auditor and her or his staff?

The audit itself is normally conducted by a team of two or more auditors. One auditor will serve as the lead in the audit, chairing the meetings, and making final decisions regarding compliance with the standard requirements. Third-party audits will begin with an opening meeting which should be attended by the organization's top level management, management representative, audit team escorts, and anyone else the company would like to include. The auditors will present the audit schedule, explain escort responsibilities, and describe how they will conduct the audit. It should be emphasized that the audit is being conducted on the system that has been established and is not intended to be an evaluation of individual personnel.

During the audit, the auditors will want to talk with personnel from all levels of management and operations. Their purpose will be to determine what actual operational practices are being followed and to compare these practices with those described in the documentation of the EMS. The documented and implemented procedures will then be evaluated to determine if the intent of the standard has been met. The auditors should not advise an organization on how it should operate; the auditors' role is to determine if the system complies with the standard. Most audits can be completed in two to four days. Each day will begin or end with a meeting between the audit team and an organization's representatives. During these meetings, the auditors will summarize any confirmed non-conformance and advise the organization of any unresolved issues.

At the conclusion of the audit, a closing meeting is held. The audit team will present findings, furnish written non-conformance statements and corrective action requests, and advise whether or not they are recommending full approval for certification, conditional approval pending completion of effective corrective action, or no approval at this time.

Each registrar has its own method for evaluating non-conformance with some utilizing the "minor and major non-conformity" format while others use a rating system based on points. Some audit programs include computerized programming in which the program not only records a text description of the finding, action, and follow-up activity in the database but also records data which may be used as measurement of performance. For every non-compliance, an action plan is established which must be completed. In one computerized program, "the action plan must be written, and must address every finding, including action on those items that were completed quickly (i.e., before the report of findings is completed). In these cases, actions must be fully described so as to document what actions were actually taken, and to allow for review of the acceptability and appropriateness of the action. Not every issue can be quickly resolved, however, and for those items, a written plan should be developed in the form of a proposal by the audited facility manager." [70] Follow-up tracking such as an Action Plan Status Report is used to identify actions completed and the status of all other pending actions.

[70] *See*, Curtis J. Johnson, "Case Study: Browning-Ferris Industries' Computerized System for Managing Audit and Environment Performance," *Total Quality Environmental Management*, Summer, 1994, 469-477.

Report and Review

Following the audit, the third-party auditing team will return to the registrar's office and begin writing the formal audit report. This report details the history of the audit, describes the strengths of the EMS system, and provides information on deficiencies found. While the auditors are completing their report, the organization should be devising and implementing corrective action for the non-conformance found during the audit; when this task is completed, they will submit to the registrar formal responses to the corrective action requests. The audit team will then review the organization's responses and make a decision regarding the efficacy of the corrective action plans. If the responses are found to effectively address the problems, the audit team will formally close the corrective action requests and recommend full approval for certification. The report, the organization's responses, and the recommendation will then be submitted to the parties making the certification decision. These decisions are made at the registrar's office which actually holds the accreditation and, in many cases, these decisions will be made in an overseas office if the United States office of the registrar is not accredited. When the certification is approved, the certificate is issued and presented to the newly-certified or registered company.

Surveillance

For third-party registration, an accredited registrar will conduct a surveillance program over the life of the certification period. Surveillance can be conducted at annual or semi-annual intervals, with most registrars conducting surveillance every six months. During the surveillance audits, auditors will carefully check the effectiveness of the internal audit program and corrective action programs in addition to reviewing several other standard elements. During the certification period, auditors will be interested in finding evidence that the EMS provides a means to achieve continuous improvement.

Most organizations will plan with their registrars for a certification program beginning with the documentation review and ending with the certificate issuance that can be completed within six to twelve months. The schedule should be planned to allow the organization time to address deficiencies as they are identified in each phase of the certification program so that certification can be achieved in the first audit.

To summarize, the following chart identifies the elements of the certification process:

Certification Process

- Application
    - Written request
    - Document Review
- Pre-Assessment (Optional)
    - Determining need for pre-assessment
    - Cursory view of operations
    - Aid in audit planning
    - Benefit to company
- Audit
    - Introductory meeting
    - Assessment
    - Closing meeting
- Report/Certification
    - Formal report issued
    - Follow-up corrective actions
    - Reassess if required
    - Certify
    - List in directory
- Surveillance

10
Legal Issues in Environmental Audits

A collision at sea can ruin your whole day.
-Thucydides

Introduction

Ensuring safety and environmental protection through an effective management system simply makes good business sense.[71] The consequences of

[71] For example, prevention of an oil spill, an actual discharge of oil, or the response to a spill may trigger, among others: the Coastal Zone Management Act; the Marine Protection, Research, and Sanctuaries Act of 1972; the Convention for the Prevention of Pollution from Ships and its Protocols (MARPOL 73/78); the Convention Relating to Intervention on the High Seas in Cases of Oil Pollution Casualties (Intervention Convention); the Protocol Relating to Intervention on the High Seas in Cases of Oil Pollution Casualties; the Federal Water Pollution Control Act and its oil pollution amendment, the Oil Pollution Act of 1990; the Refuse Act of 1899; the Deepwater Port Act; the Resource Conservation and Recovery Act; the Ports and Waterways Safety Act; the Outer Continental Shelf Lands Act; the Trans-Alaskan Pipeline Authorization Act; the International Convention on Salvage, 1989; and a myriad of state and local laws which may contradict federal and international pollution regimes. How damages to natural resources is assessed is problematic. What is the worth of a whale? The most extensively used method of valuing damage has been in terms of restoration and replacement costs; these methods do not envision the lost intrinsic value of the destroyed resource. The Oil Pollution Act of 1990 (OPA 90) provides that the National Oceanic and Atmospheric Administration (NOAA) promulgate regulations for the assessment of natural resource damages resulting from a discharge of oil. This comparatively new generation of assessment divides natural resources in terms of the direct service as well as the indirect service they provide.

Natural resources' direct services or "use services" include fishing, boating, bird watching, surfing, and research opportunities. Economic and pseudo-economic methods for valuing these include:

- Analysis of market or appraisal prices;
- Estimates of travel costs;

an oil spill,[72] for example, can be significant: Crude may be purchased for approximately 40 cents per gallon; if it were spilled, the "total spill unit value" defined as the total value for cleanup, third-party damages, and natural resource damages, averages $245.59 per gallon.[73] An effective environmental management system (EMS) may result in reduction of a company's risk of fines and/or litigation costs by timely correcting non-compliance; management of pollution control as an everyday activity rather than reacting to a crisis; achievement of cost-effective compliance with resulting reduction of cost through efficiency; retention of customer loyalty; assurance of environmental soundness to investors, thus increasing investment, loan, and planning opportunities for future growth; and reduction of insurance premiums, to name a few.

An effective EMS requires a proactive approach in which an organization ferrets out environmental problems and documents and corrects non-compliance with the plethora of laws at the federal, state, and even local levels. The daunting task of achieving full compliance, though vigorously pursued, remains elusive. Thus, it is important that environmental policy and law do not discourage this strategy; what must be encouraged is voluntary compliance, self-policing, and audits in which diligent correction of non-compliance is not penalized.

- Hedonic models; and
- Direct questioning.

Non-use services which assume an economic value to people may be: 1) vicarious consumption or option value, the knowledge that a resource exists even if use of the resource is not presently planned; 2) stewardship or existence values which includes inherent value, the benefit of a resource regardless of whether it is used, and 3) bequest values, the availability of a resource for the enjoyment of others in the future. Measurement for these are based on contingent valuation methodology (CVM). The CVM elicits a value from a responder based on her or his willingness to pay for the indirect service. To estimate the value of goods not traded on markets, a "constructed" or "hypothetical" market is devised. A survey is then conducted in which individuals are asked what they would pay to prevent a specified change in a natural resource and/or what the respondent would be willing to pay if the constructed market existed. That measurement becomes the value of that good.

[72] *See,* von Zharen, W., *Marine Transportation: Major Oil Pollution and Compensation Regimes.* 1994. In Richard Fink, editor, *Natural Resources Law Manual,* American Bar Association's Section on Natural Resources, Energy and Environmental Law, 1994, 91-113.

[73] Mercer Management Consulting carried out this study as part of its regulatory impact analysis on vessel response plans prepared for the U.S. Department of Transportation.

If the government gives special treatment to companies that conduct environmental assessments, issues arise about the audit and its findings such as whether auditing is in accordance with accepted standards and the possible resulting revelation of negligence or criminal wrongdoing or harm to public health or the environment. Privileges of protection of audit documents may limit access to relevant evidence; they are also contrary to open disclosure to stakeholders, to recent measures to encourage public disclosure, and to promotion of open, corporate environmental responsibility and stewardship. Such privileges may also run counter to the main thrust behind am EMS: Using audits as a tool to understand how to solve environmental problems.

Organizations frequently query their counsel as to what can be done to minimize environmental enforcement risk and protect business documents from discovery. There is no straight forward answer and the parameters of what is available are not easily discerned. However, several tools may be available to protect documents from discovery in litigation: the attorney-client privilege; the work-product doctrine; and the self-evaluation privilege. State "safe-harbor" laws may protect an organization from facing administrative penalties for non-compliance when such information is voluntarily and timely disclosed. Additionally, the Environmental Protection Agency has promulgated an interim environmental audit policy to encourage organizations to audit without fear of penalty.

Protection of Audits as "Privileged" Business Information

Whether a qualified evidentiary privilege for routine environmental audits is an idea whose time has come isn't clear. Fourteen states have adopted some type of qualified audit privilege. The EPA, while rejecting the privilege, has modified its policy to give organizations who audit more certainty in the realm of disclosing violations found as a result of the audit.

Corporate environmental reporting is not new. As early as the 1970s, regulations were issued by the U.S. Securities and Exchange Commission (SEC) requiring publicly-held corporations to disclose environmental compliance costs and potential liabilities to shareholders. Enforcement policies which prevent active self-policing and auditing have a chilling effect caused by the potential for involuntary disclosure. Disincentives abound: because full compliance is not possible, auditing and self-assessment will regularly uncover violations—an EMS that is proactive will find more instances of non-compliance than organizations without such systems; organizations that do disclose may be penalized more than those who are not as diligent; an effective self-policing organization who reports violations may appear to the public as a poor corporate citizen and steward of the environment; and adversaries seeking disclosure of audit documents can come in many forms: prosecutors, injured plaintiffs, civil government enforcers, citizens and citizens groups, federal EPA criminal investigators, regional EPA office enforcement specialists, municipal governments, and employee whistle-blowers.

Limiting access to or distribution of environmental audit reports and statements of environmental auditors presents a dilemma for an organization. On the one hand, company-wide dissemination of such material may lead to "lessons learned" and improvement in regulatory compliance measures. On the other hand, such documents can be powerful tools for zealous opponents in litigation. The use of privileges may protect this information. Disclosure of communications between the attorney and the client and certain other persons that relate to the attorney's legal services to the client may be prevented from disclosure under the *attorney-client privilege.* If an audit it is done under the direction and supervision of an attorney in anticipation of litigation, it may be protected by the *work-product doctrine.* Utilization of attorney-controlled protections are cumbersome, restrictive to those who would benefit, and resource-intensive. A third privilege, *self-evaluation,* may protect reports generated by self-analysis. However, its application by the courts vary widely.

The applicability of the protection privileges in environmental audits poses a different set of questions because of the unique nature of the auditing process: both in-house and outside counsel may be used; and issues which appear to be a combination of both business decisions and legal decisions may be involved and not easily separated.[74] Thus, the options for protection are limited in number and in scope.

---

[74] See *United States v. Milman,* 822 F.2d 305, 310 (2nd Cir. 1987); and *United States v. Chevron,* 757 F. Supp. 512 (E.D. Pa. 1989); see also *In Olen Properties Corp. v. Sheldahl, Inc.,* 1994 U.S. Dist. LEXIS 7125, 38 ERC 1887, 24 ELR 20936 (C.D. Cal., 12 April 1994) in which the issue of whether the attorney-client privilege is applicable to the environmental audit memoranda prepared by a person who is a client or seeking to be a client. The documents were prepared to gather information for a company's attorneys to assist the attorneys in evaluating compliance with relevant laws and regulations and appeared to have been prepared for the purpose of securing an opinion of law. The court found that these audits

Attorney-Client

The Federal Rules of Civil Procedure explicitly recognize the role of privileges as they arise under the law of evidence; one principle is certain: no privilege protects the factual information underlying an environmental audit. A legal entity such as a corporation may successfully assert the attorney-client privilege for communication with its counsel but such protection will not extend to facts: "A fact is one thing and a communication concerning that fact is an entirely different thing";[75] relevant facts cannot be concealed merely by revealing them to a lawyer. Attorney-client privilege provided to corporations is limited to communications made to attorneys solely for purpose of the corporation seeking legal advice and its counsel rendering it.[76] The attorney-client privilege was traditionally characterized as follows:

- Where legal advice of any kind is sought (as opposed to business advice);
- From a professional legal counselor in her/his legal capacity;
- The communications relating to that purpose;
- Made in confidence;
- By the client;
- Or at her/his instance and not for the purpose of committing a crime or tort is permanently protected;
- From disclosure by her/himself or by the legal adviser; and
- Except if the protection is waived.[77]

Each of these criteria can be asserted as an individual component but the criteria are not so easily reduced to a general rule of applicability. The need for

---

were privileged and need not be produced. The court looked to *United States v. United Shoe Machinery Corp.*, 89 F. Supp. 357, 35809 (D. Mass. 1950).

75 *Upjohn Co. v. United States*, 449 U.S. 383, 396, 101 S. Ct. 677, 686 (1981).

76 *In Re John Doe Corp.*, 675 F.2d 482 (2nd Cir. 1982).

77 *See* VII Wigmore, *Evidence*, Section 2292 (McNaughton Rev. Ed. 1961); *International Surplus Lines Ins. Co. v. Willis Corroon Corp.*, 1992 U.S. Dist. LEXIS 17332 (N.D. Ill. 10 November 1992) (quoting *United States v. White*, 950 F. 2d 426, 430 [7th Cir. 1991]; *In Olen Properties Corp. v. Sheldahl, Inc.*, 1994 U.S. Dist. LEXIS 7125, 38 ERC 1887, 24 ELR 20936 (C.D. Cal. 12 April 1994); *Town of Norfolk v. U.S. Army Corps of Engineers*, 968 F.2d 1438, 1457 (1st Cir. 1992).

clarification was recognized by the United States Supreme Court in the case of *Upjohn Co. v. United States*,[78] in which Upjohn, a manufacturer and seller of pharmaceuticals, conducted an audit of one of its foreign subsidiaries through an independent accountant and discovered that the subsidiary made payment to or for the benefit of foreign government officials in order to secure government business. Upjohn then initiated an extensive internal investigation in which questionnaires designated as highly confidential were sent to "all foreign general and area managers" to determine the nature and magnitude of any payments made by Upjohn or any of its subsidiaries to any employee or official of a foreign government. As part of the investigation, Upjohn's counsel also interviewed the recipients of the questionnaires as well as current and former officers and employees resulting in notes of the interviews.

When the IRS issued a summons for the production of all documents gathered during the investigation, Upjohn refused to produce them on the grounds that they were protected from disclosure by the attorney-client privilege and constituted the work-product of attorneys prepared in anticipation of litigation. A magistrate concluded that the summons should be enforced, the District Court concurred, and Upjohn appealed to the Court of Appeals for the Sixth Circuit; this court rejected the Magistrate's finding but agreed that the privilege did not apply "[to] the extent that the communications were made by officers and agents not responsible for directing Upjohn's actions in response to legal advice...for the simple reason that the communications were not the 'clients'." The Court of Appeals remanded to the District Court so that a determination could be made of who was within the "control group," that is, which employees involved in the communication were in a position of control or in a position to take a substantial part in the decision-making process or was a member of a group which had the authority to make decisions, thus making the communication privileged. Ultimately, the Sixth Circuit agreed that the documents were not protected from production since those who responded to the questionnaires were not within the control group.

The Supreme Court reversed the lower court's decision, rejecting the control group test as too narrow but refusing to undertake drafting of a set of rules which should govern challenges to investigative subpoenas. It looked to the Federal Rule of Evidence 501 which provides that "the privilege of a witness...shall be governed by the principles of common law as they may be

---

78 449 U.S. 383, 101 S. Ct. 677 (1981).

interpreted by the courts of the United States in light of reason and experience." In making its decision, the Court held that the recognition of a privilege based on a confidential relationship should be determined on a case-by-case basis. However, it did enumerate factors to be used in determining if the communications at issue are privileged:

- The communications were made in order that the corporation could obtain legal advice;
- The communications were considered "highly confidential" when made and had been kept confidential by the company;
- Communications concerned matters within the scope of the employee's corporate duties;
- The information for legal advice was not available from upper-echelon management; and
- The communications at issue were made by corporation employees to the counsel for the corporation acting as such at the direction of superiors at the corporation's order to secure legal advice from counsel.[79]

The Court failed to delineate which of the above factors must be met to trigger the attorney-client privilege. Additionally, since the Supreme Court decision interpreted a Federal Rule of Evidence, it is not precedence for state courts. Thus, one must look to individual states to ascertain each approach in identifying attorney-client privilege. It appears that a shift is toward focusing on authorization status of the employee.[80]

As noted, communication must be kept confidential. Communication to a third party will typically waive the privilege[81] including communication of

---

[79] 449 U.S. 383, 394; 101 S. Ct. 677, 685; *see also Harper & Row Publishers, Inc. v. Decker*, 423 F.2d 487 (1970) in which communication is privileged where the employee makes the communication at the direction of his superiors in the corporation and where the subject matter upon which the attorney's advice is sought by the corporation and dealt with in the communication is the performance by the employee of the duties of his employment.

[80] See *National Tank Co. v. Brotherton,* 851 S.W. 2d 193 (Tex. 1993); *Congoleum Industries v. GAF Corporation,* 49 F.R.D. 82 (E.D. Pa. 1969).

[81] See *Chandler v. Denton,* 741 F.2d 855 (Okla. 1987); *In Re John Doe Corp.,* 675 F.2d 482 (2nd Cir. 1982).

confidential material within the organization[82] if "such disclosure substantially increases the possibility that an opposing party could obtain the information disclosed."[83] However, the attorney-client privilege may cover communication to the attorney's agents. In environmental auditing, whether persons such as environmental auditors are agents of the attorney (or client) is an important question. The test is whether the communication was made in confidence and whether it was made for the purpose of obtaining legal advice from the attorney.[84] The privilege exists whether or not the person was hired by the attorney or the client as long as the advice sought is that of the attorney. Communication providing legal advice to the client involving a non-attorney employed to provide legal advice may be protected.

That an attorney was one of the people conducting the audit or to whom the results were given may not provide protection; the attorney must be acting in a legal capacity and the communication must be for the purpose of gaining or providing legal advice.[85] In *United States v. Chevron*,[86] the government pursued civil penalties for Clean Air Act violations. It sought disclosure of certain audit status reports prepared by Chevron every six months which identified a refinery's effort to correct problems during a previous environmental audit and assessed compliance with applicable laws. The audit team included senior counsel to Chevron's environmental group. Although Chevron argued that these documents would reveal the contents of the prior audit which was protected by the attorney-client privilege, the court held that the audit, including the status reports, were not privileged. Chevron had failed to show that its attorney was acting in the capacity of an attorney and that the audit's primary purpose was to obtain legal advice.

---

[82] Under some circumstances, disclosure to a third-party will not waive the privilege as when communication is disclosed pursuant to a court order. *International Business Machines Corp. v. United States,* 471 F.2d 507 (2nd Cir. 1972), *vacated on other grounds,* 480 F. 2d 293 (2nd Cir. 1973), *cert. denied,* 416 U.S. 980 (1974).

[83] *Myrna West v. Marion Laboratories,* 1991 U.S. Dist. LEXIS 18457, pp. 9-10 (W.D. Miss, 12 December 1991) (quoting *GAF Corp. v. Eastman Kodak Co.,* 85 F.R.D. 46, 52 [S.D.N.Y. 1979]).

[84] *United States v. Kovel,* 296 F.2d 918 (2d Cir. 1961).

[85] *United States v. Chevron U.S.A., Inc.,* 1989 WL 121616 (E.D. Pa. 1989).

[86] No. 88-6681, slip op. (E.D. Pa. 16 October 1989).

Merely labeling documents "attorney-client privilege" (as well as "attorney-work product") does not necessarily protect the documents.[87] The attorney-client privilege is waived when documents are treated so loosely that they should not be considered confidential and where reasonable precautions to ensure and maintain confidentiality of attorney-client documents were not made.[88] Documents passed around the offices of a corporation for review by all who care to read them will not be afforded attorney-client privilege.[89] Even if certain facts or documents were disclosed inadvertently, the protection may be waived.[90](Waiver of attorney-client privilege does not necessarily waive attorney work-product protection of an attorney's memoranda on the same subject.[91]) Thus, to trigger the privilege, strict document control must be developed and implemented. Additionally, the party claiming the privilege has the burden of showing that it has met the requirements for asserting the protection.

The communication may not be used to "sweep under the rug" past records by channeling them into the hands of an attorney:[92]

> If, in hindsight, courts determine that the audit was conducted primarily for the purpose of raising a shield against the disclosure of these occurrences by telling the lawyer about them, assuming they were not properly addressed initially, they are likely to reject assertions of the attorney-client privilege.[93]

---

87 *In Re Grand Jury,* G.J. No. 91-832 (E.D. Pa. 6 Jan 1993).

88 *O'Leary v. Purcell Co.,* 108 F.R.D. 641 (M.D.N.C. 1985).

89 *Duplan Corp. v. Deering Milliken, Inc.,* 397 F. Supp. 1146 (D.C.S.C. 1974); see also *Re LTV Secur. Litigation* 89 F.R.D. 595 (N.D. Tx. 1981) in which it was held that information gathered in a privileged manner does not become discoverable solely because the management makes other business use of such information if the information is handled in a confidential manner.

90 *Hercules Inc. v. Exxon Corp.,* 434 F. Supp. 136, 156 (D.Del. 1977).

91 *Id.*

92 *In Re International Systems & Controls Corp. Securities Litigation,* 693 F.2d 1235, 1242 (5th Cir. 1982).

93 Terrell Hunt and Timothy Wilkins, *Environmental Audits & Enforcement Policy,* 16 Harvard Env. L. Rev. 365 (1992).

For attorney-client privilege, the simple solution is to ensure steps are taken to substantiate that the legal advisor acts in that capacity and that those retained at the behest of the attorney and under the direction of an attorney (an accountant, an environmental consultant, etc.) are clearly delineated as such.[94] Practically speaking, however, environmental audit information maintained in strict confidence translates into limiting the distribution of the information as much as practically possible. Unfortunately, this is counterproductive to the spirit of self-assessment and audits in an effective EMS in which information on non-conformance and remedies should be distributed throughout the organization. Funneling information through an attorney may not provide the frank and open communication needed to find solutions to environmental problems.

Attorney Work-Product

The work-product doctrine provides another potential protection for environmental audit reports. It was recognized initially in *Hickman v. Taylor*,[95] and codified in Federal Rules of Civil Procedures.[96] The protection extends to "opinion work-product"—an attorney's conclusions, evaluations, opinions, mental impressions or legal theories; and "ordinary work-product"—all materials developed in anticipation of the trial. The basis for this is that "it is essential that a lawyer work with a certain degree of privacy."[97] Of the two

---

[94] See *Feldman v. Pioneer Petroleum, Inc.*, 87 F.R.D. 86 (W.D. Okla. 1980); see also *Burlington Industries v. Exxon Corp*, 65 F.R.D. 26 (Md. 1974).

[95] 329 U.S. 495, 67 S.Ct. 385 (1947). In this case the United States Supreme Court established three principles: 1) materials collected and prepared by counsel in anticipation of litigation are generally protected from disclosure; 2) the protection is qualified, and discovery may be required upon a sufficient showing of need by the adversary; and 3) the greatest protection is accorded materials containing the attorney's thinking, theories, analysis, mental impressions, and beliefs.

[96] "Subject to the provisions of (b)(4) of this rule, a party may obtain discovery of documents and tangible things otherwise discoverable...and prepared in anticipation of litigation or for trial by or for another party or by or for that other party's representative (including the other party's attorney, consultant, surety, indemnitor, insurer, or agent) only upon a showing that the party seeking discovery has substantial need of the materials in preparation of the party's case and that the party is unable without undue hardship to obtain the substantial equivalent of the materials by other means." Federal Rules of Civil Procedures 26 (b)(3).

[97] 329 U.S. 495, 510.

work-products, opinion work-product is given greater protection: "[F]orcing an attorney to disclose notes and memoranda of witnesses' oral statements is particularly disfavored because it tends to reveal the attorney's mental process."[98] Ordinary work-product may not be protected if an opponent is able to show there is an substantial need for the material sought in the presentation of the case or that denial of access to the information would cause an undue hardship in obtaining the equivalent of the materials by other means. The work-product doctrine requirements apply to 1) documents and tangible things; 2) prepared in anticipation of litigation or for trial; and 3) by or for another party or for that other party's representative...."[99] Corporate communications with in-house counsel which partially contain technical, non-legal matters may still be privileged if they are essentially legal in nature.[100] The attorney work-product privilege may apply to documents prepared by persons who are neither the attorney or client: The United States Supreme Court held that it is "necessary that the doctrine protect material prepared by agents for the attorney as well as those prepared by the attorney himself."[101]

Prudent management anticipates litigation and begins preparation prior to the time a suit is filed. "The test is whether, in light of the nature of the document and the factual situation in the particular case, the document can fairly be said to have been prepared or obtained because of the prospect of litigation."[102] It may be difficult to determine with precision when a particular attorney's work is prepared because of the prospect of litigation, or "some articulable claim, likely to lead to litigation"[103] rather than an investigative report developed in the ordinary course of business even when related to litigation.[104] Direct participation of an attorney as early as possible may provide evidence of materials prepared in anticipation of litigation. The attorney should expressly direct that periodic voluntary environmental audit reports be clearly identified on every page as "privileged and confidential" and

---

98 See *Hickman v. Taylor,* 329 U.S. 495, 510 (1947).

99 *Spaulding v. Denton,* 68 F.R.D. 342, 344 (D.C. Del. 1975).

100 *Schneider v. Kimberly-Clark Corp.* 91 F.R.D. 1 (N.D. Ill. 1980).

101 *United States v. Nobels,* 422 U.S. 225, 238-239, 95 S. Ct. 2160, 2170 (1975).

102 *In re Grand Jury Investigation,* 599 F. 2d 1224, 1229 (3d Cir. 1979).

103 *Coastal State Gas Corp.,* 617 F.2d at 865.

104 *Minks Mfg. Co. v. National Presto Industries, Inc.,* 709 F.2d 1109 (7th Cir. 1983); *Janicker,* 94 F.R.D. at 650; *Hoptowit v. Ray,* 682 F.2d 1237, 1212 (9th Cir. 1982); *City Consumer Serv. Inc. v. Horne,* 100 F.R.D. 740 (D. Utah 1983).

"material prepared in anticipation of litigation." This privilege can be protected only if the attorney participating in the audit carefully ensures that all audit documents remain in the realm of mental impressions, conclusions, evaluations, and opinions of the attorney. The facts on which the audit is based, however, are discoverable. The privilege is qualified: the audit report as well as the documentation developed to make the report may be discoverable if the party seeking discovery has a substantial need for the materials and such materials cannot be gathered by through other sources or by other means.

In summary, an environmental audit may occur as a routine audit without being prepared in anticipation of litigation; or it may be prepared in anticipation of litigation after an environmental incident, accident, or release. The latter would be protected. Preliminary drafts of environmental reports are protected if they are prepared in anticipation of litigation.[105] The privilege does not apply to materials prepared in the "ordinary course of business." In general, routine environmental audit documents are not privileged.

Self-Evaluative Analysis

Because organizations and individuals are called upon to make critical self-assessments for a variety of reasons, a final area of potential privilege has developed: self-evaluative analysis (also known as the self-critical analysis privilege). It is based upon the concern that disclosure of documents reflecting candid self-examination will deter or suppress socially useful investigations and evaluations or compliance with the law or with professional standards.[106] It allows individuals or businesses to:

> ...assess their compliance with regulatory and legal requirements without creating evidence that may be used against them by their opponents in future litigation. The rationale for the doctrine is that such critical self-evaluation fosters the compelling public interest in observance of the law. It protects an organization or individuals from the Hobson's choice of aggressively investigating accidents or possible regulatory violations, ascertaining the causes and results, and correcting any violations or dangerous conditions, but thereby

---

105 *Bituminous Casualty Corp. v. Tonka Corp.,* 140 F.R.D. 381 (D.Minn. 1992).

106 *Hardy v. New York News,* Inc. 114 F.R.D. 633, 640 (S.D.N.Y. 1987).

> creating a self-incriminating record that may be evidence of liability, or deliberately avoiding making a record on the subject (and possibly leaving the public exposed to danger) in order to lessen the risk of civil liability. The self-critical analysis privilege is analogous to, and based on, the same public policy considerations as Rule 407, Federal Rules of Evidence, which excludes evidence of subsequent remedial measures.[107]

The privilege of self-evaluative analysis has been successfully expounded primarily for three types of documents: hospital committee reports,[108] certain internal investigatory reports, and the various equal employment opportunity forms submitted to the government under Title VII.[109] The privilege may be applicable to any document that meets these criteria: 1) that the information is generated as a self-analysis undertaken by the party who is seeking to claim the privilege; 2) that the public must have a strong interest in preserving the free flow of the type of information sought; 3) that the information must be of the type whose flow would be curtailed if discovery were allowed; and 4) the information was prepared with the expectation that it would be kept confidential and it has in fact been kept confidential.[110] Still, the privilege "at most remains largely undefined and has not generally been recognized."[111] Several lower courts have acknowledged the privilege in other contexts,[112] but the application of self-evaluative analysis protection widely differs depending

---

107 *Reichhold Chemicals, Inc. v. Textron, Inc.*, 157 F.R.D. 522 (N.D. Fla. 1994).

108 The seminal case is *Bredice v. Doctors Hospital, Inc.*, 50 F.R.D. 249 (D.D.C. 1970).

109 See, *The Privilege of Self-Critical Analysis*, 96 Harvard Law Review 1083.

110 The fourth requirement was added by the decision in *Dowling American Hawaii Cruises, Inc.*, 971 F.2d 423, 425-26 (9th Cir. 1992).

111 *Federal Trade Commission v. TRW, Inc.*, 628 F.2d 207, 210 (D.C. Cir. 1980; *Meyers v. Uniroyal Chem. Co.*, 1992 U.S. Dist. LEXIS 6472, p.3 (E.D. Pa., 5 May 1992); *United States v. Dexter Corp.*, 132 F.R.D. 8, 9 (D. Conn. 1990).

112 *Reichhold Chemicals, Inc. v. Textron, Inc.*, 157 F.R.D. 522 (N.D. Fla. 1994). See, *e.g.*, *Dowling v. American Hawaii Cruises, Inc.*, 971 F.2d 423, 425 n.1 (9th Cir. 1992) in which voluntary routine pre-accident safety reviews are not protected by a privilege of self-critical analysis; see also *Mines Div. v. N.L.R.B.*, 805 F.2d 194, 199-200 (6th Cir. 1986) reversing an ALJ order requiring ASARCO to disclose internal self-critical reports as such disclosure would seriously affect the candor of future critiques and have a chilling effect that would defeat the critique's primary purpose.

on the court.[113] An important case to note is *Reichhold Chemicals Inc. v. Textron Inc.,*[114] a case in which Reichhold sought a protective order exempting certain documents from discovery by the privilege of self-evaluative analysis. In upholding the protection, the court noted that because pollution poses a serious public health risk and there is a strong public interest in promoting the voluntary identification and reduction of industrial pollution, "the public interest in allowing individuals and corporations to candidly assess their compliance with environmental regulations promotes sufficiently important interest to outweigh the interest of opposing private litigants in discovering this potentially highly prejudicial, but minimally relevant, evidence."[115] The court found that the company was entitled to a qualified privilege for retrospective analysis of past conduct, practices, and occurrences, and the resulting environmental consequences; and that the privilege applied only to reports which were prepared after the fact for the purpose of frank, impartial self-evaluation and analysis of the cause and effect of past pollution. The court held that evaluations of potential environmental risks of a proposed course of action, made in advance of the decision to adopt that course of action, were not protected from discovery. Thus, an organization should not depend heavily on self-evaluative protective powers in an environmental audit at this time. Nevertheless, the concept should be followed closely as more self-evaluative analyses are being contemplated.

---

[113] See, *e.g., Arkwright Mut. Ins. Co. v. National Union Fire Ins. Co. of Pittsburgh,* 1993 WL 14448 (S.D.N.Y., 11 January 1993) in which the self-evaluation privilege protected an environmental audit from discovery; see also *United States v. Dexter Corp.,* 132 F.R.D. 8 (D. Conn. 1990) in which the self-evaluative privilege was rejected in a Federal Water Pollution Control Act case.

[114] *Reichhold Chemicals, Inc. v. Textron, Inc.,* 157 F.R.D. 522 (N.D. Fla. 1994).

[115] *Id.*

Mitigation of Damages[116]

Introduction

Whether audits should be used as a shield against prosecution, or a factor to be used in mitigation of damages is a topic of compelling concern to industry. At a time when a push for environmental sensitivity is at the forefront, environmental auditing is a good "compliance" practice; its use as a shield is another question. An organization practicing due diligence is being prudent. Being prudent may be defined as effectively following ISO 14001 standards; that is, the organization has committed itself through policy and practice to protect the environment. Typically, a plaintiff must only prove that the event occurred; the defendant must prove that it did everything possible to prevent the mishap. ISO 14001 certification may demonstrate that a company has complied with environmental standards and, in turn, may receive the benefit of the doubt where necessary.

The Environmental Protection Agency's (EPA) Interim Environmental Audit Policy; Safe Harbor Laws

The EPA is responsible for obtaining compliance with federal laws that protect the environment. As such, it must by necessity enlist regulated entities to voluntarily cooperate. To this end, the EPA announced an interim environmental audit policy encouraging organizations to audit without fear of penalty. The policy is designed to provide incentives for regulated entities "that conduct voluntary compliance evaluations and also disclose and correct violations. These incentives include eliminating or substantially reducing the gravity component of civil penalties and not referring cases of criminal prosecution where specified conditions are met." [117] In some cases, the EPA will not seek criminal prosecution if the company voluntarily corrects and

---

[116] The U.S. Sentencing Guidelines established by the Sentencing Reform Act of 1984 provide criminal categories and characteristics of offenders. There is a range of sentences for each class of convicted individuals or organizations with each offense meriting a baseline number of points. It is at this point mitigating factors may reduce the number of points. Sentencing is based on the final calculation of points.

[117] *Federal Register,* 3 April 1995.

reports a violation. In most cases, it will not add punitive fines to the standard penalties for violations that are remedied and reported to the proper authorities. Specifically, major provisions include:

- Substantial reduction of civil penalties including complete elimination of the "gravity" (punitive) component of penalties for regulated entities that voluntarily identify, disclose, and correct violations according to the new EPA conditions and reduction of such penalties by up to 75% for companies that meet most, but not all of these conditions;
- That the EPA would not recommend to the Department of Justice that criminal charges be brought against a regulated entity acting in good faith to identify, disclose, and correct violations as long as no serious actual harm has occurred; and
- That the EPA will not request voluntary environmental audits of regulated entities to trigger civil or criminal enforcement investigations.

Also included is a section on the policy's relationship to emerging environmental standards such as the ISO 14000 series: "EPA also recognizes the development of and growing reliance on international voluntary environmental management standards in the U.S. and other countries. These standards, if properly crafted and implemented, can provide a powerful tool for organizations to improve their overall compliance with environmental requirements and move beyond compliance through innovative approaches to pollution prevention. In addition to issuing this interim policy, EPA will continue to pursue a dialogue with interested parties and to pilot policy approaches through programs such as the Environmental Leadership Program to determine how EPA can make use of and encourage these standards."[118]

At the same time, the EPA appears firm in maintaining its status of watchdog. The new policy holds that those states with environmental audit privilege laws—safe harbor laws—will be particularly monitored. Safe harbor laws allow for exemption from administrative penalties associated with non-compliance when an organization voluntarily discloses to an appropriate authority in a timely manner a non-compliance problem.[119] A state will be

[118] *Id.*

[119] The "Environmental Audit Protection Act" has been introduced into the Senate "to encourage owners and operators of facilities to conduct voluntary internal audits of the

especially scrutinized if it has environmental self-evaluation privileges or penalty immunities that prevent a state from obtaining:

- Information needed to establish criminal liability;
- Facts needed to establish the nature and extent of a violation;
- Appropriate penalties for imminent and substantial endangerment or serious harm to human health or the environment or from recovering economic benefit;
- Appropriate sanctions or penalties for criminal conduct and repeat violations; or
- Prompt correction of violations, and expeditious reduction of those that involve imminent and substantial endangerment to human health or the environment.

According to the Bureau of National Affairs, in 1995 some form of environmental audit privilege laws is found in: Arkansas; Colorado; Kentucky; Idaho; Indiana; Mississippi; Oregon; Utah; and Virginia. Wyoming may soon follow. Each law is different, and questions remain: Will the Department of Justice challenge these state laws? Do most state attorneys generally oppose privilege laws for audits?

Department of Justice (DOJ)

Even though the Department of Justice lauds the value of environmental auditing, it hasn't taken steps to cnsurc its confidentially. The DOJ's 1991 policy statement[120] holds that when determining whether to criminally prosecute, the DOJ attorneys should "consider" if there has been "voluntary, timely, and complete disclosure." Its "best case" for non-prosecution features "[f]ull and prompt cooperation," entailing "a willingness to make all relevant information (including the complete results of any internal or external investigation and the names of all potential witnesses) available to government

---

compliance of the facilities with federal environmental laws and for other purposes." Congressional Record, S10942-S10948 (8 August 1994).

[120] U.S. Department of Justice, *Factors in Decisions on Criminal Prosecutions for Environmental Violations in the Context of Significant Voluntary Compliance or Disclosure Efforts by the Violator* (1 July 1991).

investigators and prosecutors."[121] The policy doesn't mention protection against civil penalties.

Government attorneys, in deciding whether to prosecute criminally, will consider: 1) whether a company's audit program includes sufficient measures to identify and prevent noncompliance; and 2) whether a company has taken prompt and complete actions to correct identified problems.[122]

Conclusion and Suggestions

The ISO 14000 series requires documentation; ISO 14001 auditors trained in the ISO 9000 series format will naturally ask for documentation. It is a mistake to view an EMS system as revolving around confidentiality rather than effective management systems to protect the environment through analysis of an organization's policy, programs, activities, and their impact on the environment. Documentation is required in order to continually improve an EMS. The information in an audit report is essential to management in making sound decisions.

Commitment to environmental compliance, demonstrating an effective program to prevent and detect violations of law, and exercising due diligence to prevent and detect criminal conduct may be mitigating factors. But how does one *prove* commitment? One could envision a trial in which a variety of experts attempt to verify or deny commitment and all the while the organization is paying for legal fees. Is commitment reflected in management's attention to compliance? An integrated environmental system? Auditing? Monitoring? Training? Evaluation? Continuous improvement? It follows that if an effective EMS is used as a factor in mitigating damages, then, the potential reverse dilemma is that the absence of a system becomes a negative, aggravating element.

The EPA supports environmental audits; however, it does not support a self-evaluative privilege because of the "risk of weakening state enforcement programs, the imposition of unnecessary transaction costs and delays in enforcement actions, and the potential increase in the number of situations

---

[121] *Id.,* at pp. 3-4.

[122] U.S. Department of Justice, *Factors in Decisions on Criminal Prosecutions for Environmental Violations in the Context of Significant Voluntary Compliance,* p. 5 (1 July 1991).

requiring the expenditure of scarce Agency resources...."[123] At the same time, public disclosure of actual environmental impacts of an organization's operations, both positive and negative, may be used as a competitive advantage, using disclosure as an asset. It may force others to follow suit.

Triggering of privileges is a matter of form and proper documentation. None of the three alternatives for protection for environmental audits documents is absolute. Procedures must be in place to protect the privilege. Through careful procedural and information management in an effective EMS, privileges of protection can be maximized. Protection is most likely afforded when the documents deal with corporate legal decisions in which the attorney is significantly involved by supervising the work being done and controlling the generated documents for confidentiality. To implement this strategy for documents prepared in anticipation of litigation, an organization should document carefully:

- That audits are performed at the written request of an attorney;
- That the material is being prepared in anticipation of litigation—what the action is, why it is anticipated, and how the audit results may relate to claims;
- That an attorney is directing and supervising the documentation;
- That the attorney is acting in the capacity of legal advisor;
- That employees' responses are being used in anticipation of litigation;
- That everyone involved in the process be appraised of the necessity of keeping all communication confidential;
- That all documents be marked confidential and kept confidential (i.e., are not passed around the organization);
- That consultants are the agents of the attorney and that the contract specifies that the audit is being performed at the attorney's request to assist in legally advising the client;
- That the consultant's report is submitted to the attorney only and that the attorney will review and revise the report before it is finalized (drafts should be destroyed);
- That the attorney participates in all relevant meetings and the attendance of which is to provide legal advice to the client;

---

[123] *Restatement of Policies Related to Environmental Audits*, 59 Fed. Reg. 38455 at 38459, EPA (28 July 1994).

- That each document should be labeled "attorney-client privileged," "attorney work-product," and, where appropriate, "prepared in anticipation of litigation"; and
- That employees know that the information requested will be used by the attorney for a legal opinion.

Additionally, the following should be part of the protection strategy:

- That employees should provide information only within their scope of employment;
- That no waiver of a privilege is triggered through third-party communication;
- That there are prompt remedies for all non-compliance by setting aside in advance adequate resources to cover the reasonably anticipated costs;
- That there is put in place a process for prompt corrective action after problems are identified;
- That audit reports are written with factual data separated from evaluation.

In summary, however, where violations are identified through an audit, an organization's most prudent path is to report pertinent sections of the audit report. This results in improved regulatory relationships and exhibits a good-faith effort at being a steward of the environment.

11
Commitment to Environmental Performance and Universal Standards

The best time to take action is while still able to do so.
Admiral Halsey's comment in preparing his fleet
to meet a coming storm

ISO 14000 Recognition

The ISO 14000 series is a roadmap for environmental management in which continual improvement is sought for every environmental impact of an organization's activities. The management strategy covers the full panoply of impacts: from raw materials, design, marketing, delivery, use, and disposal of products to those of services.

The ISO 14001 specification cycle includes the development of an organization's environmental policy, identification of objectives and targets, planning, implementation and operation, audit, corrective action, and management review for continuous environmental performance improvement. It does not establish environmental performance requirements beyond what is articulated in the organization's policy, objectives and targets, and the compliance requirements of applicable legislation and regulations. It is written to apply to all types and sizes of organizations and to accommodate diverse geographical and cultural differences. The ISO 14000 series creates an environmental paradigm of thinking: of environmentally sound design and manufacturing, of product stewardship, of natural resource impacts, and of prudent resource use. The dynamic nature of an environmental management system as well as the developing processes for verifying compliance with ISO 14001 creates new opportunities for organizations attuned to environmental concerns.

Whether an organization needs a certified ISO 14001 system revolves around three points. First, the organization must be willing to fully embrace the concept of environmental management to reduce pollution and protect the environment. Secondly, a determination must be made as to whether a certified environmental management system is required to continue in or penetrate global markets. Thirdly, an organization must know its competition and its direction: an EMS may or may not be required by law; however, if competitors elect to pursue an EMS and certification, they may be perceived in the marketplace as having a higher level of environmental commitment which, in turn, could have an adverse impact on an organization that does not embrace an EMS.

Commitment

The most important phrase in all of the ISO 14000 series is "top management commitment." Commitment to effective environmental management must be anchored to the organization's overall commitment to environmental protection and continuous improvement. It should be integrated into all sections, all parts, all units throughout the organization, nation-wide and globally. To do this, the system must have active, responsible leadership with personnel at all levels involved in continuous, measurable self-assessment of the EMS.

The predominant thread in an effective environmental management system is commitment in the form of policy and goals, implementation, documentation, communication, responsibility, training, resources, and auditing. The system should include a corporate-wide policy statement emphasizing the importance of compliance with all laws to protect the environment. Sufficient resources must be available to achieve the goals. Responsibility for compliance must be designated; evidence of non-compliance must be given to appropriate personnel within the organization. There must be timely, corrective action on non-compliance issues including disclosure and reporting as required by law.

Cautionary Approach

The ISO 14000 series, as yet, should not elicit from an organization a feeling of total environmental security. Too many questions remain unanswered:

- What rules will auditors and registrars follow?
- Will occupational safety become part of the standards or must employees be trained in both conformance with ISO 14001 and in the health and safety aspects of their jobs?
- How will ISO 14001 mesh with regulatory schemes already in place, with a company's already existing management systems, and with other international or regional environmental standards?
- Will a series of auditors covering the same territory in the processes be necessary?
- Will ISO 14001 become a requirement for ISO 9002 certification?
- What employee training will be necessary?
- Will registration be interpreted as meaning compliance rather than as a starting point for continuous performance improvement?
- Will implementation and maintenance of an EMS increase costs?
- Will ISO 14001 be interpreted and applied inflexibly in the U.S.?
- Will ISO 14001 find its way into the regulatory context?
- Will the EPA adopt and require the standard?
- Will ISO 14001 be used to determine enforcement response and calculate criminal penalties?
- Will ISO 14001 increase the common law standard of care?
- What will be the driving force behind certification?

At the same time, organizations will be certified to ISO 14001 if they feel it would improve environmental performance, increase an organization's public credibility, and ultimately reduce costs both through efficiency and through the decrease liability exposure including mitigation of enforcement penalties. The standard will gain popularity if it can be integrated with other systems. Until the majority of organizations are certified, compliance with ISO 14001 may mean a competitive advantage for those who conform.

The U.K. and the European Union have developed far more prescriptive environmental standards; if they adopt the ISO 14001 standard as an equivalent, being ISO 14001 certified may become a condition of doing business in Europe. Agencies such as the World Bank and the U.S. Agency for International Development are considering requiring ISO 14001 certification as a way to assure environmental responsibility. For now, the adoption of the ISO 14000 series may be an advantage. At some point, it may be a necessity. With an effective environmental management system, an organization may come of age environmentally.

Summary

An organization should not be daunted by the complexity of the ISO 14000 series. In a nutshell, describing why an organization should care about the series is expressed by using the old journalistic stand-by:

Who? All types and sizes of businesses.

What? Organizations wanting an environmental management system that can be certifiable.

Where? For doing business around the world, and, at present, particularly in Europe.

When? Immediately although the standard is in Draft International Status.

How? Either through self-declaration or registrar.

Why? Overall, it may make good business sense; and it may soon be a requirement of doing business.

Nature is trying very hard to make us succeed; but nature does not depend on us. We are not the only experiment.

R. Buckinster Fuller, 1978.

Appendix A

ISO 14001: Environmental management systems—
Specification with guidance for use[124]

[124] *This Draft International Standard was developed by a Technical Committee of the International Organization for Standardization (ISO). The American National Standards Institute (ANSI), the U.S. Member of ISO, participates in its entire technical program and administers secretariat of 275 technical committees and subgroups. ANSI is also the exclusive sale agent in the United States for all ISO Standards, Draft International Standards, and Draft Proposals. This Table of Contents is reproduced by Dr. W.M. von Zharen through an arrangement with ANSI.*

This document is not an approved ISO International Standard. It is distributed for review and comment and may be modified during this process.

APPENDIX A

ISO 14001: Environmental management systems—Specification with guidance for use

4.3.2 Training, awareness and competence

4.3.3 Communication

4.3.4 Environmental management system documentation

4.3.5 Document control

4.3.6 Operational control

4.3.7 Emergency preparedness and response

4.4 Checking and corrective action

4.4.1 Monitoring and measurement

4.4.2 Non-conformance and corrective and preventive action

4.4.3 Records

4.4.4 Environmental management system audit

4.5 Management review

Annex A (Informative) Guidance on the use of the specification

Annex B (Informative) Bibliography

Annex C (Informative) Links between ISO 14001 and ISO 9000

Appendix B

ISO 14004: Environmental management systems—
General guidelines on principles, systems and supporting techniques[125]

[125] *This Draft International Standard was developed by a Technical Committee of the International Organization for Standardization (ISO). The American National Standards Institute (ANSI), the U.S. Member of ISO, participates in its entire technical program and administers secretariat of 275 technical committees and subgroups. ANSI is also the exclusive sale agent in the United States for all ISO Standards, Draft International Standards, and Draft Proposals. This Table of Contents is reproduced by Dr. W.M. von Zharen through an arrangement with ANSI.*

This document is not an approved ISO International Standard. It is distributed for review and comment and may be modified during this process.

APPENDIX B

ISO 14004: Environmental management systems—General guidelines on principles, systems and supporting techniques

4.2.3 Legal Requirements

4.2.4 Internal Performance Criteria

4.2.5 Environmental Objectives and Targets

4.2.6 Environmental Management Program

4.3 Implementation

4.3.1 General

4.3.2 Ensure Capability

4.3.2.1 Resources—Human, Physical, and Financial

4.3.2.2 Environmental Management System Alignment and Integration

4.3.2.3 Accountability and Responsibility

4.3.2.4 Environmental Awareness and Motivation

4.3.2.5 Knowledge, Skills and Training

4.3.3 Support Action

4.3.3.1 Communication and Reporting

4.3.3.2 Environmental Management System Documentation

4.3.3.3 Operation Control

4.3.3.4 Emergency Preparedness and Response

4.4 Measurement and Evaluation

4.4.1 General

4.4.2 Measuring and Monitoring (Ongoing Performance)

4.4.3 Corrective and Preventive Action

4.4.4 Environmental Management System Records and Information Management

4.4.5 Audit of the Environmental Management System

4.5 Review and Improvement

4.5.1 General

4.5.2 Review of the Environmental Management System

4.5.3 Continual Improvement

Annexes

A1 Examples of International Environmental Guiding Principles

A1.1 The Rio Declaration on Environment and Development

A1.2 ICC Business Charter for Sustainable Development

A2 Informative References

Appendix C

ISO 14010: Guidelines for environmental auditing—
General principles[126]

[126] *This Draft International Standard was developed by a Technical Committee of the International Organization for Standardization (ISO). The American National Standards Institute (ANSI), the U.S. Member of ISO, participates in its entire technical program and administers secretariat of 275 technical committees and subgroups. ANSI is also the exclusive sale agent in the United States for all ISO Standards, Draft International Standards, and Draft Proposals. This Table of Contents is reproduced by Dr. W.M. von Zharen through an arrangement with ANSI.*

This document is not an approved ISO International Standard. It is distributed for review and comment and may be modified during this process.

APPENDIX C

ISO 14010: Guidelines for environmental auditing—General principles

0 Introduction

1 Scope

2 Normative References

3 Definitions

4 Requirements for an Environmental Audit

5 General Principles

5.1 Objective and scope

5.2 Objectivity, independence and competence

5.3 Due professional care

5.4 Systematic procedures

5.5 Audit criteria, evidence and findings

5.6 Reliability of audit findings and conclusions

5.7 Reporting

Appendix D

ISO 14011: Guidelines for environmental auditing—Audit procedures—Auditing of environmental management systems[127]

[127] *This Draft International Standard was developed by a Technical Committee of the International Organization for Standardization (ISO). The American National Standards Institute (ANSI), the U.S. Member of ISO, participates in its entire technical program and administers secretariat of 275 technical committees and subgroups. ANSI is also the exclusive sale agent in the United States for all ISO Standards, Draft International Standards, and Draft Proposals. This Table of Contents is reproduced by Dr. W.M. von Zharen through an arrangement with ANSI.*

This document is not an approved ISO International Standard. It is distributed for review and comment and may be modified during this process.

APPENDIX D

ISO 14011: Guidelines for environmental auditing—Audit procedures—Auditing of environmental management systems

5.1.2 Preliminary document review

5.2 Preparing the Audit

5.2.1 Audit Plan

5.2.2 Audit Team Assignments

5.2.3 Working Documents

5.3 Executing the audit

5.3.1 Opening Meeting

5.3.2 Collecting Evidence

5.3.3 Audit Findings

5.3.4 Closing Meeting

5.4 Audit Reports and Records

5.4.1 Audit Report Preparation

5.4.2 Report Content

5.4.3 Report Distribution

5.4.4 Document Retention

6 Audit Completion

Appendix E

ISO 14012: Guidelines for environmental auditing—
Qualification criteria for environmental auditors[128]

[128] *This Draft International Standard was developed by a Technical Committee of the International Organization for Standardization (ISO). The American National Standards Institute (ANSI), the U.S. Member of ISO, participates in its entire technical program and administers secretariat of 275 technical committees and subgroups. ANSI is also the exclusive sale agent in the United States for all ISO Standards, Draft International Standards, and Draft Proposals. This Table of Contents is reproduced by Dr. W.M. von Zharen through an arrangement with ANSI.*

This document is not an approved ISO International Standard. It is distributed for review and comment and may be modified during this process.

APPENDIX E

ISO 14012: Guidelines for environmental auditing—Qualification criteria for environmental auditors

Appendix F

- ISM Code Elements

APPENDIX F

The following are the ISM Code Elements:

General (1.0)

Safety and environmental policy (2.0)

Company responsibilities and authority (3.0)

Designated person(s) (4.0)

Master's responsibility and authority (5.0)

Resources and personnel (6.0)

Development of plans for shipboard operations (7.0)

Emergency preparedness (8.0)

Reports and analysis of nonconformities, accidents and hazardous (9.0) occurrences

Maintenance of the ship and equipment (10.0)

Documentation (11.0)

Company verification, review and evaluation (12.0)

Certification, verification and control (13.0)

A more detailed highlight of each element includes:

ELEMENT 1. Definitions, objectives, and functional requirements.

•*Definitions:*

"Company" means the owner of the ship or any other organization or person such as the manager, or the bareboat charterer, who has assumed the responsibility for operation of the ship from the shipowner and who on assuming such responsibility has agreed to take over the duties and responsibilities imposed by the code.

"Administration" means the government of the state whose flag the ship is entitled to fly.

•*Functional Requirements For A Safety Management System (SMS)*

1. A safety and environmental protection policy.
2. Instructions and procedures to ensure safe operation of ships and protection environment in compliance with international and flag state legislation.
3. Defined levels of authority and communication between, and among, shoreside and shipboard personnel.
4. Procedures for reporting accidents and nonconformities with the provisions of the code.
5. Procedures to prepare for and respond to emergency situations.
6. Procedures for internal audits and management reviews.

•*Objectives Of The Code:*

1. Ensure safety at sea;
2. Prevention of human injury or loss of life; and
3. Avoidance of damage to the environment and property.

•*Safety Management Objectives Of The Company:*

1. Provide for safe practices in ship operation;
2. Establish safeguards against all identified;

3. Continuous improvement of safety management skills for personnel ashore and on board; and

4. Preparation for emergencies related to safety and environmental protection.

•*Safety management system should ensure:*

Compliance with mandatory rules and regulations taking into account applicable codes, guidelines and standards recommended by the organization, administrations, classification societies and maritime industry organizations.

ELEMENT No.2: Safety and environmental protection policy.

1. How to achieve stated objectives;
2. Clear and concise statements outlining philosophy of owner or operator;
3. Intent to comply with relevant regulations;
4. Encourage continuous improvement in safety practices and protection of environment;
5. Provision to review and update as necessary; and
6. Employee awareness, comprehension and execution of policies signed by CEO or senior level manager—could be separate policy statements.

ELEMENT No. 3: Company responsibilities and authority.

1. Ensure personnel ashore and on board aware of their duties and responsibilities;
2. Provide organizational chart showing those responsible for ship operations;
3. Ensure adequate resources with necessary experience and qualifications; and

4. Identify who is responsible for operation of ship if not the owner.

ELEMENT No.4: Designated person(s).

1. Designate shore personnel having direct access to the highest level of management as means for providing link between ship and shore;
2. Designees to have responsibility and authority for safety and pollution prevention aspects to vessel operation;
3. Also ensure adequate resources and shore based support provided as required; and
4. Key player(s) in ensuring SMS functions as prescribed in the code (company guidance on what correspondence is required from vessel would assist shipboard personnel e.g. minutes of safety meetings, information on injuries and oil spills).

ELEMENT 5: Master's responsibility and authority.

1. Define and document responsibility;
2. Implementing safety and environmental policies;
3. Motivating crew in observation of policy;
4. Issuing appropriate instructions and orders;
5. Verifying specified requirements observed;
6. Reviewing SMS and reporting deficiencies ashore;
7. Overriding authority on safety and pollution prevention decisions; and
8. Request company assistance as necessary (periodic visits to company office would be of benefit to both interests).

ELEMENT No. 6: Resources and personnel.

1. Master to be properly qualified, conversant with SMS and given necessary support;

2. Ship crewed with qualified and medically fit personnel (verification could be done during hiring process or through manning agent);

3. Personnel familiarization with duties; possible options are:

    (a) provide information on SMS prior to joining;

    (b) upon joining, have officer explain;

    (c) complete questionnaire;

    (d) hand-over when joining/leaving; and

    (e) video familiarization package.

4. All personnel understand rules, regulations, codes and guidelines;

5. Establish procedures for:

    (a) Identifying training;

    (b) Providing relevant information on SMS into working language of crew; and

6. Ship's personnel able to communicate effectively in execution of SMS duties.

ELEMENT No. 7: Development of plans for shipboard.

1. Company to establish procedures for preparation of plans and instructions for key shipboard operations associated with safety and pollution prevention. Examples could include (in general):

    (a) Shipboard organization:

i. Functional responsibilities;

ii. Reporting procedures;

iii. Passenger control;

iv. When applicable, communications between ship and company;

v. Inspections by Master and officers;

vi. Provision and maintenance of records;

vii. Medical arrangements, fitness avoidance of excessive fatigue;

viii. Alcohol and drug policies; and

ix. Operational and maintenance instructions for equipment.

(b) In Port Activities:

i. Cargo handling procedures;

ii. Mooring requirements;

iii. Harbor watches and patrols;

iv. Liaison with authorities;

v. Monitoring trim and stability;

vi. Procedures for repairs;

vii. Response to spill of cargo or bunkers; and

viii. Use of reception facilities for slops, noxious liquids and garbage.

(c) Preparing For Sea:

i. Verification of crew or passengers;

ii. Checking and recording draughts;

iii. Checking stability condition;

iv. Reviewing weather forecast;

v. Checking sea worthiness, securing cargo hatches, deck openings and watertight doors;

vi. Gear tests: steering gear, main engine; and

vii. Navigation and communication equipment;

(d) Harbor stations:

i. Verification of pollution prevention equipment; and

ii. Verification of publications for navigation, including charts as required by SOLAS, Chapter V, Regulation 20.

(e) The Ship at Sea:

i. Bridge and engine room watch keeping arrangements;

ii. Radio communications;

iii. Maneuvering data;

iv. Emergency procedures;

v. Fire and security patrols;

vi. Discharge overboard of oily water from machinery bilges, cargo residue from cargo tanks, noxious liquids and garbage; and

vii. Use of inert gas, where appropriate.

(f) Preparing For Arrival In Port:

i. Gear tests: steering gear, main engine, navigation and communication equipment;

ii. Harbor stations;

iii. Port information;

iv. Pilotage requirements;

v. Weather assessment;

vi. Consult publications: tide tables, coast pilot and charts; and

vii. Helicopter operations.

NOTE: The preceding lists are not exhaustive and are provided for guidance. Variations take into account circumstances, the design of the ship and particular operation and trade. The material covered does not remove the Master's authority to take whatever steps are considered necessary whether or not they are in accordance with the contents of the documentation and to issue any appropriate orders necessary for the preservation of life, the safety of the ship or the prevention of pollution.

ELEMENT No.8: Emergency preparedness

1. Company should establish procedures to identify, describe and respond to potential shipboard emergency situations as well as programs for drills and exercises;

2. Examples requiring written procedures are: steering gear failure, loss of propulsion, person overboard, loss of watertight integrity, oil spill, notification procedures following oil spill, spill containment procedures, loss of bridge control and evacuation of passengers; and

3. SMS should include measures ensuring that the company's organization can respond at any time to hazards, accidents and emergency situations involving its ships (onboard material which has relevance includes OPA-90 vessel response plan or MARPOL 26).

ELEMENT No.9: Reports and analysis of non-conformities, accidents and hazardous occurrences.

1. SMS should include procedures to ensure the following:
2. Non-conformities, accidents and hazardous situations are reported to the company;
3. All are investigated and analyzed with the objective of improving safety and pollution prevention;
4. Company to establish procedures for corrective action; and
5. Post-casualty mandatory chemical testing as dictated by flag state or company requirements.

ELEMENT No. 10: Maintenance of the ship and equipment.

1. Company should establish procedures to ensure that the ship is maintained in conformity with relevant rules and regulations. This would require:

(a) Inspections at appropriate intervals;

(b) Reporting non-conformity with cause;

(c) Corrective action to be taken;

(d) Record of all such activities;

(e) Adequate on board reference material for operation and maintenance (may include planned maintenance system, manual entry or computer program); and

(f) Procedures in SMS to identify equipment/systems which if fail, could result in hazardous situations (will vary with ship design).

ELEMENT No. 11: Documentation.

1. Company should establish and maintain procedures to control all documents and data relevant to SMS;

2. Company should ensure that:

(a) Valid documents are available where necessary;

(b) Changes to documents are reviewed and approved by authorized personnel; and

(c) Obsolete documents promptly removed.

3. Documents used to describe and implement the SMS, referred to as the "Safety Management Manual";

4. Kept in format that company considers most effective; and

5. Every ship to carry on board all relevant documentation.

NOTE: The functional requirements associated with the SMS may have already been addressed in a variety of existing acceptable documents. Therefore, a "safety management manual" which describes the SMS may include an index which directs the interested person to the substantive source document which specifically addresses the particular functional requirement under consideration.

ELEMENT No.12: Verification, review and evaluation.

1. Company or operator to carry out internal audits to verify safety and pollution prevention activities comply with the SMS;

2. Company to periodically evaluate the efficiency and review if necessary, SMS in accord with company procedures;

3. Audits and corrective actions to be carried out in accordance with documented procedures;
4. Personnel carrying out audits should be independent of the areas being audited unless impracticable due to size and nature of company;
5. Audit results and reviews should be brought to the attention of all responsible personnel; and
6. Management personnel responsible for area involved take timely corrective action on deficiencies.

ELEMENT No.13: Certification, verification and control.

1. Ship to be operated by company issued document of compliance relevant to that ship;
2. Every company complying with requirements of ISM to be issued with a document of compliance as required by the administration, or by the government of the country acting on behalf of the administration;
3. Copy of document to be available on board;
4. Safety Management Certificate to be issued to ship operating in accordance with approved SMS; and
5. Administration or organization recognized by administration should periodically verify proper functioning of the ship's SMS.

Resolutions Associated with the ISM Code.

The following provides a list and brief description of those resolutions associated with the provisions of ISM:

- *Principles of Safe Manning. A481(12).* This was adopted on 19 November 1991. All Administrations were recommended to provide each of their registered ships with a document specifying the minimum number and grades of qualified seafaring personnel required to be carried.

- *Control by the Flag State Over the Owner of a Ship. A441(11).* This was adopted on 15 November 1979. The intent was to enhance the ability of the State to exercise control over the owner of a ship flying its flag or over the person contracted or otherwise entrusted by the owner of such ship to discharge her/his responsibilities in these matters.

- *General Principles for Ship Reporting Systems and Ship Reporting Requirements including Guidelines for Reporting Incidents Involving Dangerous Goods, Harmful Substances, and/or Marine Pollutants." A648(16).* This was adopted on 19 October 1989, in order to ensure that ship reporting systems and reporting requirements comply as closely as possible with the general principles specified in the Annex to the present resolution. Additionally, States having been advised of an incident may take such measures on the high seas as may be necessary to prevent, mitigate, or eliminate grave and imminent danger to their coastline or related interests from pollution or threat of pollution of the sea by oil and substances other than oil following a maritime casualty.

Conventions Associated with the ISM Code.

The following provides a list and brief description of those conventions associated with the provisions of ISM:

- *International Convention for the Safety of Life at Sea, 1974: SOLAS 74.* This Convention provides a comprehensive range of minimum standards for the safe construction of ships and for the basic safety equipment (i.e., fire prevention, navigational, life-saving and radio) to be carried onboard. SOLAS 74 also contains operational instructions, particularly on emergency procedures, and provides for regular surveys and for the issuance of certificates of compliance. The three basic certifications are:

1. Safety of Life at Sea - Construction (SLC);
2. Safety of Life at Sea - Equipment (SLE); and
3. Safety of Life at Sea - Radiotelegraph/radio Telephone (also known as GMDSS [Global Maritime Distress Safety System]).

- *The International Bulk Carrier (IBC) and the International Gas Carrier (IGC) Codes, Mandatory Requirements under SOLAS 74.* The former is for the transportation of dangerous chemicals in bulk and the latter is for the transportation of dangerous gases in bulk. Certification and periodic surveys are required to attest and confirm vessel compliance.

• *International Convention for the Prevention of Pollution from Ships as modified by the protocol of 1978. MARPOL 73/78.* This Convention, more thoroughly discussed in section B2c, contains measures designed to prevent pollution caused both accidentally and in the course of routine operations. Five annexes in the convention cover pollution by oil, noxious liquid substances in bulk, harmful substances carried in packaged forms, sewage and garbage.

• *Convention on the International Regulations for Preventing Collisions at Sea, 1972. COLREG.* These "rules of the road" provide basic provisions for rights of way, safe speed, action to avoid collision, and procedures to observe in narrow channels and in restricted visibility. Also included are navigation light requirements.

• *International Convention on Load Lines, 1966.* This Convention sets the minimum permissible freeboard according to the season of the year and the trading area of the ship. It also contains special ship construction standards with regard to water tightness.

• *Merchant Shipping (Minimum Standards) Convention. International Labor Organization (ILO) Convention 147.* This Convention requires Administrations to have effective legislation on safe crewing standards, hours of work, seafarers' competency, and social security. It also sets employment standards equivalent to those contained in a range of ILO instruments covering, for example, minimum age, medical care and examination, social security, and training.

- *International Convention on Standards of Training, Certification, and Watchkeeping for Seafarers, 1978. STCW 78.* This Convention provides training, certification and qualification requirements for: senior officers; all officers in charge of watches in the deck, engine and radio departments; and rating forming part of a watch. All such seafarers are required to have a certificate, endorsed in a uniform manner. It also specifies basic principles to be observed in keeping deck and engine watches and special qualification requirements for personnel on oil, chemical, and liquefied gas tankers.

Deficiencies.

It is impracticable to define a ship as substandard solely by reference to a list of qualifying defects. Having assessed the extent with which operational requirements are complied, the control officer then has to exercise her/his professional judgment to determine whether the operational proficiency of the crew as a whole is of a sufficient level to allow the ship to sail without a major risk for the safety of life at sea or threat to the marine environment, or whether a better level of proficiency should be required. For this latter purpose, following the procedures set out in resolution A481(12), the ship may be detained.

Appendix G

- ISMA Code Elements

APPENDIX G

The ISMA Code contains twenty-two elements or chapters which encompass all aspects of vessel operation. Unlike ISM, it addresses the areas of insurance and finance. The chapters include:

General: Scope, Objectives, References, and Definitions (1.)

Business Ethics (2.)

Organization (3.)

Personnel (4.)

Safety (5.)

Environmental Protection (6.)

Contingency Planning (7.)

Operational Capability (8.)

Cost Efficiency/ Purchasing/Contracting (9.)

Maintenance/Maintenance/Standard (10.)

Technical Support (11.)

Insurance (12.)

Accounting (13.)

Certification and Compliance with Rules and Regulations (14.)

Cargo Handling and Cargo Care (15.)

Communication Procedures (16.)

Management Agreement (17.)

Records (18.)

Auditing Body (19.)

Quality System (20.)

Document Control (21.)

Internal Quality Audits (22.)

Appendix H

Rio Declaration on Environment and Development

The Rio Declaration On Environment and Development

The United Nations Conference on Environment and Development, having met at Rio de Janeiro form 3 to 14 June 1992, reaffirming the Declaration of the United Nations Conference on the Human Environment, adopted at Stockholm on 16 June 1972, and seeking to build upon it, with the goal of establishing a new and equitable global partnership through the creation of new levels of cooperation among States, key sectors of societies and people, working towards international agreements which respect the interests of all and protect the integrity of the global environmental and developmental system, recognizing the integral and interdependent nature of the Earth, our home proclaims that:

Principle 1:

Human beings are at the center of concerns for sustainable development. They are entitled to a healthy and productive life in harmony with nature.

Principle 2:

States have, in accordance with the Charter of the United Nations and the principles of intentional law, the sovereign right to exploit their own resources pursuant to their own environmental and developmental policies, and the responsibility to ensure that activities within their jurisdiction or control do not cause damage to the environment of other States or of areas beyond the limits of national jurisdiction.

Principle 3:

The right to development must be fulfilled so as to equitably meet developmental and environmental needs of present an future generations.

Principle 4:

In order to achieve sustainable development, environmental protection shall constitute an integral part of the development process and cannot be considered in isolation from it.

Principle 5:

All States and all people shall cooperate in the essential task of eradicating poverty as an indispensable requirement for sustainable development, in order to decrease the disparities in standards of living and better meet the needs of the majority of the people of the world.

Principle 6:

The special situation and needs of developing countries, particularly the least developed and those most environmentally vulnerable, shall be given special priority. International actions in the field of environment and development should also address the interests and needs of all countries.

Principle 7:

States shall cooperate in a spirit of global partnership to conserve, protect and restore the health and integrity of the Earth's ecosystem. In view of the different contributions to global environmental degradation, States have common but differentiated responsibilities. The developed countries acknowledge the responsibility that they bear in the international pursuit of sustainable development in view of the pressures their societies place on the global environment and of the technologies and financial resources they command.

Principle 8:

To achieve sustainable development and a higher quality of life for all people, States should reduce an delineate unsustainable patterns of production and consumption and promote appropriate demographic policies.

Principle 9:

States should cooperate to strengthen endogenous capacity-building for sustainable development by improving scientific understanding through exchanges of scientific and technological knowledge, and by enhancing the development, adaptation, diffusion and transfer of technologies, including new and innovative technologies.

Principle 10:

Environmental issues are best handled with the participation of all concerned citizens, at the relevant level. At the national level, each individual shall have appropriate access to information concerning the environment that is held by public authorities, including information on hazardous materials and activities in their communities, and the opportunity to participate in decision making processes. States shall facilitate and encourage public awareness and participation by making information widely available. Effective access to judicial and administrative proceedings, including redress and remedy, shall be provided.

Principle 11:

States shall enact effective environmental legislation. Environmental standards, management objectives and priorities should reflect the environmental and developmental context to which they apply. Standards applied by some countries can be inappropriate and of unwarranted economic and social cost to other countries, in particular developing countries.

Principle 12:

States should cooperate to promote a supportive and open international economic system that would lead to economic growth and sustainable development in all countries, to better address the problems of environmental degradation. Trade policy measures for environmental purposes should not constitute a means of arbitrary or unjustifiable discrimination or a disguised restriction on international trade. Unilateral actions to deal with environmental challenges outside the jurisdiction of the importing country should be avoided. Environmental measures addressing transboundary or global environmental problems should, as far as possible, be based on an international consensus.

Principle 13:

States shall develop national laws regarding liability and compensation for the victims of pollution and other environmental damage. States shall also cooperate in an expeditious and more determined manner to develop further international law regarding liability and compensation for adverse effects of environmental damage caused by activities within their jurisdiction or control to areas beyond their jurisdiction.

Principle 14:

States should effectively cooperate to discourage or prevent the relocation and transfer to other States of any activities and substances that cause severe environmental degradation or are found to be harmful to human health.

Principle 15:

In order to protect the environment, the precautionary approach shall be widely applied by States according to their capabilities. Where there are threats of serious or irreversible damage, lack of full scientific certainty shall not be used as a reason for postponing cost-effective measures to prevent environmental degradation.

Principle 16:

National authorities should endeavor to promote the internalization of environmental costs and the use of economic instruments, taking into account the approach that the polluter should, in principle, bear the cost of pollution, with due regard to the public interest and without distorting international trade and investment.

Principle 17:

Environmental impact assessment, as a national instrument, shall be undertaken for proposed activities that are likely to have significant adverse impact on the environment and are subject to a decision of a competent national authority.

Principle 18:

States shall immediately notify other States of any natural disasters or other emergencies that are likely to produce sudden harmful effects on the environment of those States. Every effort shall be made by the international community to help States so afflicted.

Principle 19:

States shall provide prior and timely notification and relevant information to potentially affected States on activities that can have a significant adverse transboundary environmental effect and shall consult with those States at an early stage and in good faith.

Principle 20:

Women have a vital role in environmental management and development. Their full participation is therefore essential to achieve sustainable development.

Principle 21:

The creativity, ideals and courage of the youth of the world should be mobilized to forge a global partnership in order to achieve sustainable development and ensure a better future for all.

Principle 22:

Indigenous people and their communities, and other local communities, have a vital role in environmental management and development because of their knowledge and traditional practices. States should recognize and duly support their identity, culture and interest and enable their effective participation in the achievement of sustainable development.

Principle 23:

The environment and natural resources of people under oppression, domination and occupation shall be protected.

Principle 24:

Warfare is inherently destructive of sustainable development. States shall therefore respect international law providing protection for the environment in times of armed conflict an cooperate in its further development, as necessary.

Principle 25:

Peace, development and environmental protection are interdependent and indivisible.

Principle 26:

States shall resolve all their environmental disputes peaceful and be appropriate means in accordance with the Charter of the United Nations.

Principle 27:

States and people shall cooperate in good faith and in a spirit of partnership in the fulfillment of the principles embodied in this Declaration and in the further development of international law in the field of sustainable development.

Appendix I

The ICC Business Charter for Sustainable Development

International Chamber of Commerce (ICC) Business Charter for Sustainable Development

1. Corporate Priority

   To recognize environmental management as among the highest corporate priorities and as a key determinant to sustainable development; to establish policies, programs and practices for conducting operations in an environmentally sound manner.

2 Integrated Management

   To integrate these policies, programs and practices fueling into each business as an essential element of management in all its functions.

3. Process of Improvement

   To continue to improve policies, programs and environmental performance, taking into account technical developments, scientific understanding, consumer needs and community expectations, with legal regulations as starting point, an to apply the same environmental criteria internationally.

4. Employee Education

   To educate, train and motivate employees to conduct their activities in an environmentally responsible manner.

5. Prior Assessment

To assess environmental impacts before starting a new activity or project and before decommissioning a facility or leaving the site.

6. Products or Services

To develop and provide products or services that have no undue environmental impact and are safe in their intended use, that are efficient in their consumption of energy and natural resources, and that can be recycled, reused, or disposed of safely.

7. Customer Advice

To advise, and where relevant, educate customers, distributors, and the public in the safe use, transportation, storage and disposal of products provided; and to apply similar considerations to the provisions of services.

8. Facilities and Operations

To develop, design and operate facilities and conduct activities taking into consideration the efficient use of energy and materials, the sustainable use of renewable resources, the minimization of adverse environmental impact an waste generation, and the safe and responsible disposal of residual wastes.

9. Research

To conduct or support research on the environmental impacts of raw materials, products, processes, emission, and wastes associated with the enterprise and on the means of minimizing such adverse impacts.

10. Precautionary Approach

To modify the manufacture, marketing, or use of products or services or the conduct of activities, consistent with scientific and technical understanding, to prevent serious or irreversible environmental degradation.

11. Contractors and Suppliers

To promote the adoption of these principles by contractors acting on behalf of the enterprise, encouraging and, where appropriate, requiring improvements in their practices to make them consistent with those of the enterprise; and to encourage the wider adoption of these principles by suppliers.

12. Emergency preparedness

To develop and maintain, where significant hazards exist, emergency preparedness plans in conjunction with the emergency services, relevant authorities and the local communicate, recognizing potential transboundary impacts.

13. Transfer of Technology

To contribute to the transfer of environmentally sound technology and management methods throughout the industrial and public sectors.

14. Contributing to the Common Effect

To contribute to the development of public policy and to business, governmental and intergovernmental programs and education initiatives that will enhance environmental awareness and protection.

15. Openness to Concerns

To foster openness and dialogue with employees and the public, anticipating and responding to their concerns about potential hazards and impacts to operations, products, wastes or services, including those of transboundary or global significance.

16 Compliance and Reporting

To measure environmental performance; to conduct regular environmental audits an assessments of compliance with company requirements, legal requirements and these principles and periodically to provide appropriate information to the Board of Directors, shareholders, employees, the authorities and the public.

Appendix J

ACRONYMS

APPENDIX J

ANSI	American National Standards Institute. Adopts but does not write American standards. Assures that member organizations that do write standards follow rules of consensus and broad participation by interested parties. ANSI is the US member of ISO.
ASQC	American Society for Quality Control, an administrator of the US TAG to TC 207.
ASTM	American Society for Testing and Materials. a US standards-making body that has developed a standard on EMS evaluation. It is an administrator of the US TAG to TC 207.
CEN	Comite Europeen de Normalisation (European Committee for Standardization)
CENELEC	Comite Europeen de Normalization Electrotechnique (European Committee for Electrotechnical Standardization)
CD	Committee Draft
CMA	Chemical Manufacturers Association
CSA	Canadian Standards Association
DFE	Design for the Environment
DIS	Draft International Standard
DOD	United States Department of Defense
DOE	United States Department of Energy

EA	Environmental Audit
EAC	European Accreditation of Certification
EAF	Environmental Auditing Forum
EAPS	Environmental Aspects in Product Standards
EAR	Environmental Auditing Roundtable
EC	European Commission
EEC	The European Economic Community. This encompasses the EC and EFTA countries.
EFTA	European Free Trade Association. A group of nations whose goal is to remove import duties, quotas and other obstacles to trade and to uphold nondiscriminatory practices in world trade. Current members are Austria, Finland, Iceland, Norway, Sweden and Switzerland.
EL	Environmental Labeling
EMAR	Eco-Management and Audit Regulation (EU)
EMAS	Eco-Management and Audit Scheme (EU)
EMS	Environmental Management System
EPA	United States Environmental Protection Agency
EPE	Environmental Performance Evaluation
EPI	Environmental Performance Indicators
EU	European Union. The EU is a framework within which member states have agreed to integrate their economies and eventually form a political union. Current members are Austria, Belgium, Denmark, Finland, France, Germany,

	Greece, Ireland, Italy, Luxembourg, Netherlands, Portugal, Spain, Sweden, and the United Kingdom.
GATT	General Agreement on Tariffs and Trade
GEMI	Global Environmental Management Initiative, a Washington, DC based company that deals with international business and environmental issues.
IAAR	Independent Association of Accredited Registrars
ICC	International Chamber of Commerce. An organization based in Paris, France, participating in the drive for international environmental standardization.
ICSP	Inter-Agency Committee on Standards Policy (within NIST)
IEA	Institute of Environmental Auditing
IEC	International Electrotechnical Commission. an equivalent of ISO
IESU	International Environmental Systems Update
IQA	Institute for Quality Assurance
ISO	International Organization for Standardization. A worldwide federation of national standards bodies (111 at present). ISO produces standards in all fields except electrical and electronic (which are covered by IEC). Formed in 1947.
LCA	Life-Cycle Assessment (or Life-Cycle Analysis)
MRA	Mutual Recognition Agreement.
NACCB	National Accreditation Council for Certification Bodies (UK). The British authority for recognizing the competence

	and reliability of organizations that perform third-party certification of products and registration of quality systems. Formed in 1984.
NCSCI	National Center for Standards and Certification Information (within NIST)
NIST	National Institute of Standards and Technology. Department of Commerce
NSFI	NSF International. a US standards-making body that has developed its NSF 110 EMS standard, similar to ISO 14001.
NTIS	National Technical Information Service
OSHA	United States Occupational Safety and Health Administration
RAB	Registrar Accreditation Board. A US organization whose mission is to recognize the competence and reliability of registrars of quality systems, and to achieve international recognition of registrations issued by accredited registrars. It is seeming to be the US's ISO 14001 accreditation body.
RvC	Raad voor de Certificatie (Dutch Council for Certification). The Dutch authority for recognizing the competence and reliability of organizations that perform third-party certification of products, accreditation of laboratories and registration of quality systems. The first such organization, formed in 1980.

SAGE	Strategic Advisory Group on Environment, formed by ISO to assess the need for an international EMS standard--since disbanded.
SC	Subcommittee
SCRAG	Standards Conformance Registrar Advisory Group, formed by the US TAG to develop criteria for a IJS accreditation body--since disbanded.
SIC	Standard Industrial Classification code
SME	Small and Medium-Sized Enterprises
ST	SubTAG. Subgroup of a TAG
SubTAG	Subgroup of a TAG
SWG	Subworking Group (US TAG)
TAG	Technical Advisory Group. A term used in the United States for groups that are responsible for input on international standards within their respective scopes; other countries may use other terms.
TC	Technical Committee, as in TC 207
TG	Task Group
WD	Working Draft
WG	Working Group

Appendix K

GLOSSARY

APPENDIX K

Accreditation	Procedure by which an authoritative body formally recognizes that a body or person is competent to carry out specific tasks.
Ancillary Material	Material input that is used by the system producing the product. but is not used directly in the formation of the product.
Applicant	Legal entity applying for an environmental label for a product or a range of products and which undertakes the compliance with ecological and product function criteria and the certification and costs involved in the application and awarding of the label.
Assessment	An estimate or determination of the significance. importance, or value of something.
Assessment Body	Third party that assesses products and registers the quality systems of suppliers.
Assessment System	Procedural and managerial rules for conducting an assessment leading to issue of a certification document and its maintenance.

Audit	A planned, independent and documented assessment to determine whether agreed-upon requirements are being met.
Audit Conclusion	Professional judgment or opinion expressed by an auditor about the subject matter of the audit, based on and limited to reasoning the auditor has applied to audit findings.
Audit Criteria	Policies, practices, procedures or requirements against which the auditor compares collected evidence about the subject matter.
Audit Findings	Result of the evaluation of the collected audit evidence compared against the agreed audit criteria.
Audit Program	The organizational structure, commitment and documented methods used to plan and perform audits.
Audit Team	Group of auditors, or a single auditor. designated to perform a given audit. The audit team may also include technical experts and auditors in training. One of the auditors on the audit team performs the function of lead auditor.
Auditee	Organization to be audited.
Auditor	Individual performing an environmental audit, or part thereof, who meets the criteria specified in ISO 14012.

Certification	Procedure by which a third party gives written assurance that a product, process, or service conforms to specified requirements.
Certified	The EMS of a company, location or plant is certified for compliance to ISO 14001 after it has demonstrated such compliance through the audit process. When used to indicate EMS certification, it means the same thing as registration.
Client	Organization commissioning the audit. The client may be the auditee, or any other organization which has the regulatory or contractual right to commission an audit.
Comparative Assertion	Environment claim regarding the superiority of one product versus a competing product.
Compliance	An affirmative indication or judgment that the supplier of a product or service has met the requirements of the relevant specifications, contract, or regulation; also the state of meeting the requirements.
Conformance	An affirmative indication or judgment that a product or service has met the requirements of the relevant specifications, contract or regulation; also the state of meeting the requirements.

Conformity Assessment	Conformity assessment includes all activities that are intended to assure the conformity of products to a set of standards. This can include testing, inspection, certification, quality system assessment and other activities.
Continual Improvement	Process of enhancing the environmental management system, with the purpose of achieving improvements in overall environmental performance, not necessarily in all areas of activity simultaneously, resulting from continuous efforts to improve in line with the organization's environmental policy.
Contractor	The organization that provides a product to the customer in a contractual situation.
Convention	A customary practice, rule or method.
Corrective Action	An action taken to eliminate the causes of an existing nonconformity, defect or other undesirable situation in order to prevent recurrence.
Customer	Ultimate consumer, user, client, beneficiary, or second party.
Data Quality	Degree of confidence in individual input data from a source, aggregated data and in the data set as a whole. Life Cycle Assessment data quality is

	described by established indicators selected for the study.
Design Review	A formal, documented, comprehensive and systematic examination of a design to evaluate the design requirements and the capability of the design to meet these requirements and to identify problems and propose solutions.
Elementary Flow	Any flow of raw material entering the system being studied and which has been drawn from the environment, that is without previous human transformation; any flow of material leaving the system being studied, and which is discarded into the environment, that is without subsequent human transformation.
EN 45000	A series of standards set up by the EC to regulate and harmonize certification, accreditation- and testing activities. Guides for assessment and accreditation of certification bodies and guides for bodies operating product certification systems are expected to be published by the end of 1995.
Environment	Surroundings in which an organization operates, including air, water, land, natural resources, flora, fauna. humans, and their interrelation. The

	environment in this context extends from within an organization to the global system.
Environmental Aspects	Elements of an organization's activities, products, and services which can interact with the environment.
Environmental Audit	Systematic, documented verification process of objectively obtaining and evaluating evidence to determine whether specified environmental activities, events, conditions, management systems, or information about these matters conform with audit criteria, and communicating the results of this process to the client.
Environmental Impact	Any change to the environment, whether adverse or beneficial, wholly or partially resulting from an organization's activities, products, or services.
Environmental Impact	Consequences for human health, for the well-being of flora and fauna or for the future availability of natural resources, attributable to the input and output streams of a system.
Environmental Labeling-Type I	Multiple criteria-based, third-party voluntary environmental labeling program.

Environmental Management	Parts of the overall management function of an organization that develop, implement, achieve, review, and maintain the environmental policy.
Environmental Management System	EMS—organizational structure. responsibilities, practices, procedures, processes, and resources for implementing and maintaining environmental management.
EMS Audit	A systematic and documented verification process to objectively obtain and evaluate evidence to determine whether an organization's environmental management system conforms to the EMS audit criteria set by the organization, and to communicate the results of this process to management.
EMS Audit Criteria	Requirements derived from policies, practices, procedures and other elements, as covered by ISO 14001 and, if applicable, any additional EMS requirements against which the auditor compares collected evidence about the subject matter.
Environmental Objectives	Overall environmental goals, arising from the environmental policy and significant aspects that an

	organization sets itself to achieve, and which are quantified wherever practicable.
Environmental Performance	The measurable outputs of the environmental management system, relating to an organization's control of the impact of its activities, products, or services on the environment, based on its environmental policy, objectives, and targets.
Environmental Policy	Statement by the organization of its intentions and principles in relation to its overall environmental performance, which provides a framework for action and for the setting of its environmental objectives and targets.
Environmental Targets	Detailed performance requirements, quantified wherever practicable, applicable to the organization or parts thereof, that arise from the environmental objectives and that need to be set and met in order to achieve those objectives.
Evidence	Verifiable information, records or statements of fact.
Finding	A conclusion of importance based on observation(s).
Follow-up Audit	An audit whose purpose and scope are limited to verifying that corrective action has been

	accomplished as scheduled and to determining that the action effectively prevented recurrence.
Functional Unit	Measure of performance of the main functional outputs of the product or service system.
Impact Assessment	Phase of Life Cycle Assessment aimed at understanding and evaluating the magnitude and significance of environmental impacts based on the life cycle inventory analysis.
Inspection	Activities such as measuring, examining, testing, gauging one or more characteristics of a product or service and comparing these with specified requirements to determine conformity.
Interested Parties	Individuals or groups concerned with or affected by the environmental performance of an organization.
Inventory Analysis	Phase of Life Cycle Assessment involving the compilation and quantification of inputs and outputs for a given product or service system throughout its life cycle.
Joint Assessment	Cooperative assessments resulting in formal mutual recognition of certifications.
Lead Auditor	Auditor leading a specific environmental audit, who meets the criteria specified in ISO 14012.

Life-Cycle	Consecutive and inter-linked stages of a product or service system, from the extraction of natural resources to the final disposal.
Life-Cycle Assessment	Systematic set of procedures for compiling and examining the inputs and outputs of materials and energy and the associated environmental impacts directly attributable to the functioning of a product or service system throughout its life-cycle.
Non-Renewable Resource	Natural resource that cannot be replaced, regenerated or brought back to its original state once it has been extracted.
Nonconformity	The nonfulfillment of a specified requirement.
Notified Body	A notified body is a testing organization that has been selected to perform assessment activities for (a) particular directive(s). It is approved by the competent authority of its member state and notified to the European Commission and all other member states.
Organization	Company, operation, firm, enterprise, institution, or association, or part thereof, whether incorporated or not, public or private.

Organizational Structure	The responsibilities, authorities and relationships, arranged in a pattern, through which an organization performs its functions.
Practitioner	Private or governmental sponsored entity which develops, owns, sponsors and/or administers an environmental labeling program.
Prevention of Pollution	Use of processes, practices. materials, products or energy that avoid or reduce the creation of pollution and waste.
Procedure	A specified way to perform an activity.
Process	A set of interrelated resources and activities which transform inputs into outputs.
Product	Goods and services for consumer, commercial, and industrial purposes.
Product Criteria	Set of qualitative and quantitative technical requirements that the applicant, product, or product category shall meet to be awarded an environmental label. Product criteria include ecological and product function elements.
Protocol Agreement	An agreement signed between two organizations that operate in different but complementary fields of activity and that commit themselves to take into account their respective assessment results accordingly to conditions specified in advance.

Purchaser	The customer in a contractual situation.
Qualification Process	The process of demonstrating whether an entity is capable of fulfilling specified requirements.
Raw Material	Primary or secondary recovered or recycled material that is used in a system to produce a product.
Recycling	Set of processes for reclaiming material that would otherwise be disposed of as a material input to a product or service system.
Registration	Procedure by which a body indicates relevant characteristics of a product, process or service or particulars of a body or person, and then includes or registers the product. Process or service in an appropriate publicly available list.
Renewable Resource	Natural resource that is capable of regeneration.
Requirements of Society	Requirements including laws, statutes, rules and regulations, codes, environmental considerations, health and safety factors, and conservation of energy and materials.
Resource Depletion	Reduction in the global stock of raw material as a result of extraction of non-renewable resources, or extraction of renewable resources faster than they can be renewed.

Responsible Care	Comprehensive guidelines for environmental management systems adopted by Chemical Manufacturers Association (CMA) in 1988. Participation by individual businesses is an obligation of membership in the CMA.
Root Cause	A fundamental deficiency that results in a nonconformance and must be corrected to prevent recurrence of the same or similar nonconformance.
Service	The result generated by activities at the interface between the supplier and the customer and by supplier internal activities to meet the customer needs.
Service Delivery	Those supplier activities necessary to provide the service.
Service System	Collection of materially and energetically connected operations which performs one or more defined functions.
Specification	The document that prescribes the requirements with which the product or service must conform.
Stakeholders	Those groups and organizations having an interest or stake in a company's EMS program (*e.g.*, regulators, shareholders, customers, suppliers, special interest groups, residents, competitors, investors, bankers, media, lawyers, insurance

	companies, trade groups, unions, ecosystems, cultural heritage, and geology).
Subcontractor	An organization that provides a product to the supplier.
Supplier	An organization that provides a product to the customer.
Survey	An examination for some specific purpose: to inspect or consider carefully: to review in detail.
System Boundary	Interface between the product or service system being studied and its environment or other systems.
Technical Expert	Individual who provides specific knowledge or expertise to the audit team, but who does not participate as an auditor.
Testing	A means of determining an item's capability to meet specified requirements by subjecting them to a set of physical, chemical, environmental, or operating actions and conditions.
Third Party	Person or legal entity recognized as being independent of the parties involved in the sale of a product. The practitioner, or its agent, is a third-party. Suppliers or producers are the first party and consumers the second party.

Valuation	Element of impact assessment in which the relative significance of the data from the inventory analysis or results of the characterization are established.
Verification	Process of authenticating evidence. The act of reviewing, inspecting, testing, checking, auditing or otherwise establishing and documenting whether items, processes, services or documents conform to specified requirements.
Waste	Any output from the product or service system which is disposed of.

Index

D

E

F

G

H

I

Q

R

S

Design for the Environment: Product Life Cycle Design Guidance Manual
This new life cycle approach examines each step in the life of a product from acquisition of raw materials through processing, manufacture, use, and final disposal of all residuals.
Softcover, 181 pages, Mar '94, ISBN: 0-86587-384-4 **$65**

Environmental Science and Technology Handbook
This is the first book to bridge the gap between the latest environmental science and technology available, and compliance with today's complex regulations.
Hardcover, 389 pages, Jan '94, ISBN: 0-86587-362-3 **$79**

Bioremediation of Hazardous Waste: EPA Research Series
By Environmental Protection Agency
Compiled from four EPA documents, this book contains over 35 papers discussing approaches to the bioremediation of hazardous waste sites.
Softcover, 188 pages, Mar '94, ISBN: 0-86587-388-7 **$55**

Total Quality for Safety and Health Professionals
F. David Pierce, a CSP and a CIH, shows you how to apply concepts - proven successful - to your safety management program to achieve increased productivity, lowered costs, reduced inventories, improved quality, increased profits, and raised employee morale.
Hardcover, 244 pages, June '95, ISBN: 0-86587-462-X **$59**

Environmental Regulatory Glossary, 6th Edition
This glossary defines and standardizes more than 4,000 terms, abbreviations and acronyms, compiled directly from the environmental statutes or the U.S. Code of Federal Regulations.
Hardcover, 544 pages, June '93, ISBN: 0-86587-353-4 **$68**

Pollution Prevention Strategies and Technologies
This book is an indispensible guide to understanding pollution prevention policies and regulatory initiatives designed to reduce wastes.
Hardcover, Index, 484 pages, Oct '95, ISBN: 0-86587-480-8 **$79**

South Korea Environmental Report: A Resource for Business
This new, authoritative report provides you with a wealth of information on environmental regulatory programs in South Korea.
Softcover, Index, 260 pages, Feb '96, ISBN: 0-86587-508-1 **$495**

Environmental Statutes, 1996 Edition
The complete text of each statute as currently amended is included, with a detailed Table of Contents for quick reference.
Hardcover, 1,200 pages, Mar '96, ISBN: 0-86587-521-9 **$69**
Softcover, 1,200 pages, Mar '96, ISBN: 0-86587-522-7 **$59**
Disks include the Folio search Engine.
3.5" Floppy Disks for Windows (#4058) **$135**
Statutes Package, hardcover w/Windows disk (#4059) **$204**

Federal Facility Pollution Prevention: Planning Guide and Tools for Compliance
This EPA Guide presents pollution prevention tools and provides a step-by-step approach to develop a pollution prevention program plan for federal facilities.
Softcover, 240 pages, Aug '95 ISBN: 0-86587-476-X **$75**

Bioremediation of Ground Water and Geological Material: A Review of In-Situ Technologies
By Environmental Protection Agency
This EPA manual provides a detailed background of the technologies available, as well as the most recent scientific understanding of the processes involved for the bioremediation of contaminated soil and ground water.
Softcover, 250 pages, June '94, ISBN: 0-86587-404-2 $75